Implementing Integrated Water Resources Management in Central Asia

NATO Science Series

A Series presenting the results of scientific meetings supported under the NATO Science Programme.

The Series is published by IOS Press, Amsterdam, and Springer in conjunction with the NATO Public Diplomacy Division.

Sub-Series

I. Life and Behavioural Sciences	IOS Press
II. Mathematics, Physics and Chemistry	Springer
III. Computer and Systems Science	IOS Press
IV. Earth and Environmental Sciences	Springer

The NATO Science Series continues the series of books published formerly as the NATO ASI Series.

The NATO Science Programme offers support for collaboration in civil science between scientists of countries of the Euro-Atlantic Partnership Council. The types of scientific meeting generally supported are "Advanced Study Institutes" and "Advanced Research Workshops," and the NATO Science Series collects together the results of these meetings. The meetings are co-organized by scientists from NATO countries and scientists from NATO's Partner countries—countries of the CIS and Central and Eastern Europe.

Advanced Study Institutes are high-level tutorial courses offering in-depth study of latest advances in a field.
Advanced Research Workshops are expert meetings aimed at critical assessment of a field, and identification of directions for future action.

As a consequence of the restructuring of the NATO Science Programme in 1999, the NATO Science Series was re-organized to the four sub-series noted above. Please consult the following web sites for information on previous volumes published in the Series.

http://www.nato.int/science
http://www.springer.com
http://www.iospress.nl

Series IV: Earth and Environmental – Vol. 77

Implementing Integrated Water Resources Management in Central Asia

edited by

Patricia Wouters
University of Dundee, U.K.

Victor Dukhovny
Scientific Information Center of Interstate Coordination
Water Commission Tashkent, Uzbekistan

and

Andrew Allan
University of Dundee, U.K.

Springer

Published in cooperation with NATO Public Diplomacy Division

Proceedings of the NATO Advanced Research Workshop on
Integrated Resources Management in Transboundary Basins – An
Interstate and Intersectoral Approach
Bishkek, Kyrgyzstan
23–28 February 2004

A C.I.P. Catalogue record for this book is available from the Library of Congress.

ISBN 978-1-4020-5731-1 (PB)
ISBN 978-1-4020-5730-4 (HB)
ISBN 978-1-4020-5732-8 (e-book)

Published by Springer,
P.O. Box 17, 3300 AA Dordrecht, The Netherlands.

www.springer.com

Printed on acid-free paper

CONTENTS

FOREWORD

"*Without water there is no life*"

This book is the result of a collaborative project between the UNESCO Centre for Water Law, Policy and Science (formerly the International Water Law Research Institute) at the University of Dundee, and the Scientific Information Center of the Interstate Commission for Water Coordination in the Aral Sea basin (SIC ICWC). The project was dedicated to examining the practical issues of Integrated Water Resource Management (IWRM) in Central Asia, and culminated in a workshop held at Bishkek in February 2004. The workshop brought together some of the leading experts in the field and resulted in important insights into some of the issues surrounding the transboundary and national management of the region's shared fresh water. However, the speed of developments in the region has been so rapid that even in the relatively short time since the workshop took place, significant changes have taken place.

As a consequence of the disintegration of the Soviet Union and the resulting emergence of five newly independent states in Central Asia—Kazakhstan, Kyrgyzstan, Tajikistan, Turkmenistan, and Uzbekistan—the water resources that in the past were the sovereign domain of a single country (USSR) had become transboundary. This precipitated the need for finding new approaches to manage the now transboundary waters of the two main rivers of the region—the Amudarya and Syrdarya Rivers—in order to meet national demands of the five sovereign States that depended upon these waters. Thus, in September 1991, 1 month after independence, the national Ministers of water resources from each of the five countries concluded and signed a document setting forth the agreed need for joint water management of the basin waters. On February 18, 1992, an agreement was reached on the procedures for such management and the Interstate Commission for Water Coordination (ICWC) was established as the management agency responsible for the joint management of the transboundary waters in the region.

From its beginnings in the early 1990s, the international cooperation of the Central Asian countries in the field of water resources has further developed and matured, primarily through the efforts of the ICWC as a focal coordinating joint body. Through years of cooperative endeavors the approaches to water resources management adopted in the region have emerged as relatively unique in many ways. There are very few other examples from the world's 250 transboundary basins where water resource planning and regulation takes place at the interstate level, and where continuous efforts related to the allocation of water resources are undertaken on a regular basis. Following is a list of some of the major achievements accomplished over the course of recent years:

- Three international agreements between the Central Asian States, signed by the Heads of State, and two framework agreements on water management, were adopted during the last decade demonstrating the political will to engage in collaboration aimed at achieving a conflict-free water allocation between the parties.
- The ICWC has maintained continuous water management, planning, and control with respect to the utilization of transboundary rivers through both its operational bodies and the mechanism of regular (quarterly) meetings of the parties. In addition to addressing annual and seasonal water allocation matters, these meetings consider strategic concerns, including the improvement of management systems, implementation of information and data monitoring systems, integrated management of water resources, and so forth. In total, 43 meetings of the ICWC have been held.
- Despite extreme weather variations over the last few years, including two particularly humid and three exceptionally dry years, the ICWC has succeeded in preventing conflicts relating to water management, allocation, and use through extensive and well-coordinated actions. As a result of these efforts the total annual water withdrawals in the region have been reduced from 110 to 103 km^3.
- The regional Information System and information portal that integrate systematic data from both national water institutions and basin water-management organizations were established with the support of the European Union and the Swiss Development Agency.
- The countries of the region have established effective advanced capacity-development systems in the form of the regional Training Centre and its national branches. Over the last 5 years, more than 2000 medium- and higher-level experts and specialists from water-management institutions and water user associations received training. In addition, 2500 farmers receive annual training through a network of demonstration plots.
- The so-called "integrated water resources management" approach (IWRM) was recognized as the most appropriate method of ensuring sustainable water use and development in the region in the context of growing water scarcity. Accordingly a wide range of specific measures for rational water use have been implemented at the national level. The implementation of IWRM in the three selected irrigation systems in Kyrgyzstan, Tajikistan, and Uzbekistan demonstrated the effectiveness and practicability of this approach.

However, as evidenced by the joint activity progress reports and outside observers, some of the ICWC activities, especially those related to the elaboration of regional water policy and further improvement of relevant legal frameworks, have not advanced as much and as quickly as was initially hoped. These concerns were addressed in detail at the workshop in Bishkek, which brought together all ICWC members, including leaders of the five national water departments, managers of the two river basin organizations, as well as prominent international and local experts. Their papers published in this book analyze both the current state of the water management of transboundary water resources in the Aral Sea basin, its strengths and weaknesses, and discuss ways to further improve the overall effectiveness of transboundary regional cooperation in the field.

The opening paper by Professor Tarlock provides a general review of IWRM principles and assesses their evolution from the perspective of current world development trends. These trends, in his opinion, are formed by a shift in the understanding of water resources as public property to a perception of water as a factor in sustainable development at the intersection of competing interests.

It should be noted in this respect that historically the Central Asian tradition gave priority to nature in water demand management and in determining the limits of acceptable water withdrawals from rivers. This tradition also imposed strict limitations on water use and distribution. Public water management in the context of the Islamic Sharia law included both appointment of water managers by the public, public participation in all water activities, water distribution rules "from root water users to upper levels" and public arbitration of all water disputes, all of which are features of the modern IWRM.

Current efforts to reach an appropriate balance between, for example, commercial energy interests and irrigated agriculture face almost impossible challenges in the context of the existing upward trends in energy prices, along with the decline of agricultural production prices (exacerbated by the destructive impact of agricultural subsidies in developed countries). Consequently, different approaches are needed achieve a proper balance between competing demands and here the unique experience of the USA–Canada International Joint Commission is instructive.

The emphasis of modern transboundary water law on the principle of equitable and reasonable use causes problems of implementation because of the lack of precise criteria implicit in the concept. New approaches to the application of equitable and reasonable use have recently emerged as demonstrated by the Legal Assessment Model (LAM), developed by the water resource experts from the University of Dundee. Other tools to address water sustainability issues have also been put forward in the Central Asian region, including those proposed by V. Prikhodko and I. Rusiev, which link national socioeconomic development and scenario forecasting.

The papers of the water ministers from the region reflect specific approaches adopted by each country with respect to its water development agenda. All of them, however, show an appreciation of the importance of jointly agreed rules and procedures in the Aral Sea basin. Some essential aspects of the river basin management are highlighted in various papers. The report by Zh. B. Bekbolotov (Kyrgyzstan), as well as the paper by Sh. Mukhamedjanov (SIC ICWC) underline the significance of the main objective of IWRM, that is improvement of water productivity along with the use of water reserves, and highlight the application of water charges as an efficient economic tool in achieving this objective. The underestimation of the importance of drainage (A. A. Djalalov, Uzbekistan) also has a considerable impact on water productivity.

Almost all countries of the region (papers by A. A. Djalalov, A. A. Nazirov, A. Sh. Djaloobayev) consider the transition to hydrographic method of water management and the establishment of WUA as first steps toward IWRM. However, it should be remembered that the hydrographic method alone, without public participation, is fraught with the risk of replacing the "administrative hydro-egoism" of upstream users with "professional hydro-egoism", i.e., a complete lack of public supervision of the activities of water management institutions. The hydrographic approach combined with public participation and vertical interlinking of all levels in the administrative hierarchy are demonstrated in the first large scale project of IWRM in the Fergana Valley (a joint project involving the Swiss Agency for Development and Cooperation, the International Water Management Institute, and the SIC ICWC) (paper by A. A. Nazirov et al.).

It would be unwise, however, to think that the introduction of IWRM is limited only to institutional, legal, and financial reforms of water management. As demonstrated by the experience gained in implementing the Fergana Valley project, technical and managerial tools play an important role in improving management efficiency (papers by A. I. Tuchin, M. Kh. Khamidov).

Minister A. A. Nazirov (Tajikistan) analyses in his paper a range of key issues that must be addressed in developing an appropriate legal framework for water resources management in the region:

- consideration of long-term multipurpose reservoir regimes and their interrelationship with the interests of the energy sector, irrigation, and the environment;
- specific operational procedures of the Basin Water Organizations (BWOs) relating to situations of water scarcity and floods;

- a need to establish a Water-Energy Consortium as a region-wide economic and financial mechanism to coordinate the operational regimes set up by the ICWC and the energy demand interests of upstream countries;
- a need for long-term agreements between riparian states on both the Amudarya and the Syrdarya rivers, that take account of their future development.

Minister T. A. Altiyev (Turkmenistan) addresses the role of strategic and national planning in water development aimed at meeting the interests of riparian countries concerned. Using his country as an example, he looks at the unique experience of the world's largest canal, the Karakum.

Serious environmental and socioeconomic problems of the lower reaches of a transboundary river is dealt with in a paper by Dr. T. K. Kamalov, the former leader of Karakalpakstan—the most water stressed zone in the entire Aral Sea basin. In his opinion, the implementation of IWRM at the basin level should ensure a sustainable water supply to this environmentally and socially disadvantaged part of the basin. The mechanisms for implementation of such management are discussed in detail in the report of Yu. Khudaiberganov, chief of the BWO "Amudarya," a basin that has experienced particular difficulties as a result of two consecutive dry years.

The fundamental principles of the BWO operation are presented in the paper by M. Kh. Khamidov, chief of the BWO "Syrdarya." These principles form the basis for attaining sustainable development of transboundary water resources in the region through the implementation of IWRM at the basin level by providing for:

- integration of national legal frameworks with regional agreements;
- mechanism for reaching consensus during negotiations;
- application of economic mechanisms;
- consideration of environmental requirements;
- further enhancement of the independent status of regional water organizations.

More than 2 years have passed since the workshop in Bishkek and considerable progress has been made toward the introduction and practical implementation of IWRM in the Aral Sea basin. The IWRM-Fergana Project has successfully proceeded to its third phase, which anticipates a wider uptake of this approach in the Fergana Valley and in other projects as well; the justification for the project "IWRM in lowlands" was made on the basis of the former. With the help of the Asian Development Bank, the ICWC has moved forward in elaborating general approaches to the revision of the interstate agreements on the Syrdarya River and on a range of other draft agreements. The spirit of collaboration is being strengthened, and this gives confidence that IWRM will actually be developed and properly implemented in the Aral Sea basin as a whole.

The University of Dundee UNESCO Centre for Water Law, Policy and Science now works with the region to enhance the capacity of water resources experts through the implementation of the innovative "Water Law, Water Leaders" program. This is designed to foster the creation of a new generation of local water leaders through the development and delivery of UK postgraduate degrees at the local level. The idea of "taking the Masters to the river" is aimed at encouraging regional cooperation through the establishment of regional knowledge hubs. This approach is unique and demonstrates the foresight and vision emanating from the region. With the global water challenge growing more complex every day, it is encouraging to see the progress being made in Central Asia. For more information on the Water Law, Water Leaders program, see www.dundee.ac.uk/water.

The editors wish to gratefully acknowledge the financial assistance received from the NATO Science Program, without which this meeting would not have taken place. Thanks are also due to Gordon Spark of the University of Dundee, whose assistance in the editorial process was invaluable.

INTEGRATED WATER RESOURCES MANAGEMENT—INTERNATIONAL BEST PRACTICE

INTEGRATED WATER RESOURCES MANAGEMENT: THEORY AND PRACTICE

A. Dan Tarlock
Distinguished Professor of Law, Chicago-Kent College of Law, Chicago, IL 60661-3691, USA

I. INTRODUCTION: ENVIRONMENTALLY SUSTAINABLE RIVER USE: A NEW PARADIGM

1. From Commodity to Dynamic Ecosystem

Water resource planning and management are undergoing a paradigm shift. Historically, rivers have been viewed as communities to be exploited to the maximum extent possible for economic development. Water resource planning has primarily been an engineering exercise to achieve the optimum development of river basins for hydropower, flood control and consumptive use. Throughout the world, countries have constructed large-scale multiple-purpose dams and irrigation systems. Both international and domestic water law has supported optimum development by (1) creating semi-exclusive national rights to divert and store water and (2) and encouraging unilateral national water resources development. Water management meant the enforcement of existing entitlements and adherence to the original project purposes. The traditional vision of a river system of a commodity to be put to the optimum or maximum use remains the dominant paradigm in many parts of the world such as China, Central Asia, India and other areas as a matter of choice or necessity. However, the traditional paradigm is slowly being replaced by the alternative paradigm of ecologically sustainable development (ESD).

The basic reason for the paradigm shift is that multiple-purpose development has imposed substantial environmental costs and increased social inequity in many river basins. In recent years, there have been many efforts to measure the historically ignored costs of intensive river basin development.[1] The process of environmental accounting has led to a more radical ecological ideal of managing river systems to maintain some measure of a river's historic ecological services,[2] such as those provided by floods and seasonable flow variations, as well as to support a wide range of consumptive uses in a more efficient, sustainable manner.

P. Wouters et al. (eds.), Implementing Integrated Water Resources Management in Central Asia, 3–21.

The newer ecological integrity vision is less clearly articulated than multiple-use development because it rests on a more complex view of the human role in the functioning of natural systems.[3] It starts from the premise that we must try to integrate human uses of a river system with the maintenance of its natural environmental sustainability,[4] both in the design of new projects and the re-engineering and operation of existing facilities.[5] It is not a simple river preservation concept but rather tries to adapt the emerging international norm of environmentally sustainable development (ESD) to water use and management. For example, the 1998 Report of the United States Western Water Policy Review Advisory Commission adopted a following of sustainable water use:

> Sustainable water resource management builds on the long tradition of state and federal management to conserve water and apply it to a wide range of beneficial uses, but the achievement of sustainability also presents new challenges for which past management practices and institutions often provide limited guidance.[6]

The Commission identified the core element of a sustainable water policy as follows: we need to define hydrologic baselines for individual basins and watersheds that reflect the full range of valued waster uses, including ecosystem uses.[7] Thus, environmentally sustainable water use is not a simple river preservation concept. Although some aquatic scientists want to subordinate human use to the normative river,[8] most visions of environmentally sustainable water management recognize the need to maintain and sometimes expand consumptive uses. ESD requires water resource regimes which strike a new balance between in and out-of-stream uses and reduce the economically irrational subsidies that have characterized much water use in water-stressed countries.

There is a high degree of recognition students of water policy and large segments of water use community that future water policy will be guided by three substantive principles[9]: (1) the more efficient use of existing supplies, (2) the use of more sophisticated, less environmentally intrusive technologies to develop new supplies, and (3) the restoration of degraded aquatic ecosystems to maintain and recapture valuable ecosystem services.[10] In addition, there is a widespread consensus that water resource planning should not simply be closed dialogue among experts but should be more inclusive. Participation in the planning and management of water should be expanded to include diverse stakeholders, user communities, cultural minorities and others often excluded from access to these decisions and supporting information in the past.[11] These objectives require countries to adopt new planning and management strategies and to support them with more effective national water laws and international agreements.

2. The Factors Driving the Paradigm Shift

Three primary changed conditions drive the paradigm shift. The first changed condition is the growing consensus that the earth's climate is warming and that this warming can distort normal water allocation patterns. The precise impacts on specific basins and sub-basins are difficult to predict because the climate change models suggest that the impacts of climate change will vary greatly among the earth's regions. Any watershed or river basin prediction must deal with high levels of hydrologic, economic and political uncertainty. However, the general risks that arid areas face can be stated with some confidence, although the geographic and temporal scale of the change is uncertain.

Some regions, such as sub-Saharan Africa, may experience decreased precipitation and more extended droughts.[12] Areas with present abundant supplies such as the Mekong basin may face new conflicts because of the combination of population pressure and a decreased annual run off. Other regions will see increased precipitation and more frequent and more severe floods. Increased precipitation is not necessarily a blessing because it may not translate into more available water supplies in all regions. In water-short areas with historically variable rainfall patterns, increased precipitation may actually exacerbate efforts to provide reliable water supplies. More precipitation may fall as winter rain rather than snow, and thus the snowpacks may melt earlier as warmer average temperatures mean that Spring runoffs will come earlier and evaporate faster. In addition, nations and regions may have to adapt to series ecosystem changes and these must be factored into any adaptation strategy. However, this will be difficult at the present time because the state of climate change research does not permit managers to go from large-scale models to specific basins and from basin models to regional predictions.

The second changed condition is the winding down of the era of large dam and water project construction. The 2000 publication of the Report of the World Commission on Dams recommended a more rigorous assessment of proposed new dams and that much attention be focused on the re-operation of existing dams and irrigation systems and on the promotion of more sustainable water storage and use technologies. The Report adopts, in large measure, the environmental and social critique of large dams of large dams, especially in developing countries. For example, it found that irrigation dams have typically short of physical targets, did not recover their costs and have been less profitable in economic terms,[13] and they displace large populations. The story is more dismal for ecosystem effects. Dams inundate large areas and kill terrestrial plants and displace animals; reservoirs may account for between 1% and 28% of all green house gas emissions. Large dams compromise the dynamic aspects of rivers that is fundamental to maintaining the character of aquatic ecosystems.[14]

The third changed condition is the growing pressure to improve or restore degraded aquatic ecosystems. A series of influential studies in the United States, Europe and the Middle East[15] has recently led to the radical idea of managing river systems to maximize ecological functions. The newer ecological integrity vision is less clearly articulated than multiple uses because it rests on a more complex view of the human role in the functioning of natural systems, including floods. It starts from the premise that we must try to integrate human uses of a river system with the maintenance of its natural environmental sustainability both in the design of new projects and the re-engineering existing projects. The current focus is on river restoration because so many large systems have been modified. The newly developing science of conservation biology furnishes the scientific underpinnings for the vision. In brief, all river systems-modified and natural must be seen as dynamic, ever-changing functioning ecosystems which serve a variety of functions from the maintenance of consumptive uses to the provision of valuable ecosystem services.[16]

II. INTEGRATED WATER RESOURCES MANAGEMENT

1. History of Concept

The current thinking in the world water community is that ESD should be implemented by Integrated Water Resource Management (IWRM). IWRM is not a totally new concept. Nor, is it a substitute for making hard political choices about alternative water use options. There is a long history of treating river basins systems as single units and trying to plan and execute comprehensive management regimes characterized by integrated dams and irrigated systems. IWRM builds on this long tradition of river basin planning, with deep roots in the United States and the former Soviet Union, but tries to correct the environmental and social myopia of previous planning and water resources development models as well as to introduce greater public involvement and economic discipline into water management and allocation practice. It focuses much more on developing a wider range of alternatives to achieve long-term environmentally and socially sustainable water uses compared to previous planning models and practices. IWRM is river basin or catchment area-focused, and thus it can be confined to the national level or expanded across national boundaries.

IWRM was endorsed in Agenda 21, the environmental action plan for the 21st century agreed to at the 1992 United Nations Rio de Janeiro Conference on Environment and Development (UNCED). It is also one of the six principles adopted at the 1992 Dublin Conference on Water and the Environment.

Between UNCED in 1992 and the 2002 World Summit on Sustainable Development (WSSD) or Rio Plus 10 in Johannesburg, South Africa, IWRM was endorsed by the Commission on Sustainable Development,[17] the General Assembly of the United Nations,[18] and the Ministerial Declaration of the International Conference on Freshwater.[19]

IWRM as the international water management norm was reaffirmed at the World Summit on Sustainable Development (WSSD), held in Johannesburg in 2002. The WSSD re-articulated that ESD consists of three mutually reinforcing principles: economic development, social development and environmental protection.[20] WSSD's primary contribution is to add social development as a separate and coequal principle. Previously, it was only a component of economic development. In developing IWRM, the WSSD emphasizes the extent to which human needs are fulfilled by environmental protection only to the extent that it is an integral part of economic and social development. In short, IWRM has always been based upon the foundational premises of ESD, but the WSSD re-emphasizes the extent to which IWRM must now integrate economic and social development with environmental protection. In short, IWRM, counsels that new demands for water must be recognized as potential constraints on traditional, often inefficient uses to accommodate new environmental and social equity uses.

2. IWRM Objectives

IWRM calls for the holistic management of freshwater as a finite and vulnerable resource, and the integration of sectoral water plans and programs within the framework of economic and social policy.[21] The objectives of integrated water resources management, as articulated in Agenda 21, are:

1. The promotion of a dynamic, interactive, iterative and multisectoral approach to water resource management, including the identification and protection of potential sources of freshwater supply, that integrates technological, socioeconomic, environmental and human health considerations;
2. The planning of strategies for the sustainable and rational utilization, protection, conservation and management of water resources based on community needs and priorities within the framework of national economic development policy;
3. The design, implement and evaluate projects and programmes that are both economically efficient and socially appropriate within clearly defined strategies, based on an approach of full public participation, including that of women, youth, indigenous people and local communities in water management policymaking and decision making;

4. The identification strengthening, or development, as required, in particular in developing countries, the appropriate institutional, legal and financial mechanisms to ensure that water policy and its implementation are catalysts for sustainable social progress and economic growth.

III. IWRM IN PRACTICE

Many countries of the world have adopted IWRM in principle. This is not surprising. Very few countries want to endorse formally unsustainable water use and ad hoc, uncoordinated water planning, although both are widely practiced. IWRM is also often a way to legitimate new, controversial, objectives. For example, Brazil's new water policy adopts IWRM. The country did so, in part, to justify bulk water tariffs to support watershed and basin planning and to finance the administration of water permits.[22] In other cases, the adoption of IWRM is part of the project of entrance into the world (western) community of nations. China's new water law creates a framework to achieve IWRM.[23] The 2000 European Union Water Framework Directive adopts IWRM to improve the water quality of the Union's heavily used rivers. The Directive requires a river basin management plan that prioritizes risks and establishes cost effective measures to reduce pollution loads and flood damage.[24]

IWRM is easy to adopt in principle, but far fewer countries have put IWRM into practice. I have selected four examples of national water laws and policies that can serve as models of IWRM. Viable IWRM models must have the following characteristics: (1) the holistic treatment of water resources, (2) the elevation of environmental protection to parity with consumptive use, (3) the use of inclusive, continuing adaptive planning processes, (4) the restriction of present and future consumptive uses, and (5) the recognition of equity or social justice as constraints on water allocation. Not all countries exhibit all of these standards, but each country in the process of implementing one of more of them. In addition, as will be discussed in Section III, IWRM must be supported by effective national or transnational water laws that in fact limit unsustainable uses of water.

1. South Africa

South Africa illustrates a regime that elevates environmental protection to an equal position with consumptive uses, recognizes equity or social justice claims and constrains wasteful consumptive uses. Post-apartheid South Africa adopted a very progressive water code which provides for both environmental protection and social equity.[25] The Code redressed the previous bias toward large diversions for irrigated agriculture, which resulted in the denial of adequate water access to much of the country's majority black population. The 1998 Water Act adapted the United States doctrine that the use of water is subject to the public trust[26] to create two water reserves. The reserves both guarantee water to satisfy basic human needs and to protect aquatic ecosystems in order to secure ecologically sustainable development and the use of the relevant water resource.[27] A quantified reserve will be established for each covered water resource. In 2002, the government released the Proposed First Edition of the National Water Resource Strategy which surveyed 19 water management areas and concluded that between 12 and 30 percent of the river's base flow will be required to maintain the ecological reserves.[28]

2. The Murray—Darling Basin in Australia

Australia's management of the Murray—Darling Basin, which is plagued with the whole range of environmental problems from saline land to degraded aquatic ecosystems, is one of the world's most ambitious ecosystem revival experiments because it seeks to promote a new management regime in a basin that is and will be used for intensive irrigated agriculture. It is an example of an effort to establish a normative river. River ecologists have proposed this term to describe management regimes that use pre-demand diversion baselines.[29] It also illustrates how an environmental standard can be used to mitigate previous harmful resource use decisions, and thus limit existing entitlements throughout this system.

In 1992, the federal government and the basin's four federal states (New South Wales, Queensland, South Australia and Victoria) agreed to the Murray—Darling Initiative to conserve the river's ecosystem and thus address serious land salinization throughout the basin. The Initiative led to the adoption of the federal-state Murray-Basin Agreement[30] and the creation of a joint federal-state commission overseen by a federal-state ministerial council. Unlike a United States interstate water allocation compact or an international allocation treaty, the agreement imposes much more detailed land use and water management duties on the parties and is constantly amended by new agreements. It both allocates the flow among the basin states[31] and vests the Commission with the power to control releases from specified upstream storage facilities. The Murray—Darling Commission now runs the river, overseen by the ministerial council and a stakeholder advisory board, although the four federal states must ultimately decide how individual water entitlements are adjusted to basin-wide objectives.[32]

The most important potential international precedent is the Commission's adoption of an artificial base flow regime and the imposition of the regime on existing users throughout the basin. The Commission has initiated a process to set environmental or base flows for ecosystem restoration based on the impacts of different flows on the riverine environment. On developed river basins, the problem with establishing new flow or hydrograph regimes is vested or claimed rights. Both the federal and state governments recognized the need to limit water withdrawals, establish base flows and to stabilize and restore productive agricultural areas, especially those degraded by salinization. In 1995, the Commission announced a "cap" which is the "*cornerstone of a number of policies designed to manage water resources for scarcity: water trading, environmental flows and the security of property rights.*"[33] The cap imposes yearly diversion limits on the four basin states and the Australian Capital Territory and thus will vary from year to year according to the supply. New South Wales agriculture accounts for many of the stresses on the system, and the Commission imposed a state cap based on 1993–94 withdrawal levels.[34]

Australia is prone to prolonged period of severe droughts which alternate with wet years. Diversions are increasing upstream in Queensland as well in many of New South Wales inland agricultural districts. In 1996–1997 three major sub-basins in New South Wales exceeded the cap.[35] Achievement of the cap will require many innovative, aggressive management strategies such as conjunctive use of ground and surface water, an abandonment of the "use it or lose" it administration of water licenses and the implementation of an accounting system to balance water use over a period of time. Still, the Commission predicted that the cap would be met in all states except New South Wales. Only one basin, the Lachlan, clearly exceeded the cap in its first years of implementation, 1997–98, but usage in other major basins is approaching the cap, especially if development is allowed to increase.[36] Thus, the ability of the cap to help restore the Murray-Darling will not be known for years.[37] In late 2003, the Commission's Scientific Review Panel issued a controversial interim report which concluded that 1500 giga liters (GLs) of recovered environmental flows per year would be necessary to restore the health of the Murray. However, the Ministerial Council issued only a first step decision which set a goal of only 500 GLs per year and postponed implementation of an inter-state water sharing plan until mid-2004.[38]

The initial experience suggests plans which first try to maintain the status quo and then promote gradual and modest rollbacks in existing uses can be both fair, efficient and promote environmental objectives. In major river systems, there are almost always wasteful agricultural water use and use in excess of legal entitlements so river managers have some flexibility to experiment without the undue dislocation of legitimate user expectations.

The ultimate test of IWRM is whether it drives subsequent resource choices, including the reduction of existing claimed entitlements. There are various ways to do this, and the cap induced a significant flexibility device to do this, the Pilot Interstate Trading program in the Mallee Region of South Australia, Victoria and New South Wales along the lower Murray. Both water prices and agricultural crops are similar among the three states. Under the Pilot Program, individual diverters with high security water rights may sell water across state lines, provided that the water licensing authorities in each state agree to the transfer.[39] One of the major unresolved issues in water marketing is how to integrate the benefits of markets with environmental protection objectives. The Murray-Darling Pilot Program does this by establishing exchange values (the amount of water that can actually be transferred) among states. Trades by upstream diverters from New South Wales to Victoria and from Victoria to South Australia have a 1.0 exchange rate which means that 100% of the entitlement can be transferred. But, transfers from South Australia, which is downstream, to the upstream states of Victoria and New South Wales have an exchange rate of 0.9 so that only 90% of the entitlement can be transferred to prevent further salt concentrations in the river.

3. The Great Lakes

The Canada-United States Great Lakes are an example of a transnational water management regime that it is trying to implement many of the principles of IWRM without formally adopting the concept. The two countries have come to view the lakes as a single ecosystem that should be conserved through the coordinated review of diversions that threaten to disturb the dynamic, long-term stability of the system. Threats exist because the amount of fresh water in them makes the lakes a prime candidate, at least in the eyes of many in Canada and the United States, for trans-basin diversions to augment supplies in water-short areas. There have been proposals to transfer water from Lake Superior to the High Plains of the central United States and to withdraw water for bulk transfer to water-short regions in other parts of the world. In addition, global climate change helps fuel the persistent regional fears that the lakes will be tapped to augment water supplies outside the basin. On one level, the lakes are a classic example of an underdeveloped property rights regime. There is an extensive, but primarily soft legal regime that governs the lakes but neither the United States nor Canada have allocated the right to use the lakes between the two countries. No interstate allocation has been made among the United States basin states so there are no clearly recognized entitlements to use Great Lakes water. However, there is an inchoate "Law of the Lakes" with substitutes a process of state, provincial and international study and consultation for a hard allocation regime.

The most interesting feature of this evolving regime is the preference it accords non-consumptive over consumptive uses. The soft law of the Great

Lakes gives considerably more weight to the conservation of the lakes' ecological services compared to other allocation regimes. The littoral states and the two federal governments have evolved a weak legal regime to protect the most important regional component of the lakes, the maintenance of naturally fluctuating levels, that can be the basis for adapting to global climate change. This regime has minimized conflicts by limiting and discouraging consumptive use, although it has also retarded the development of a firmer property rights regime for the lakes and formal inter-basin cooperative planning structures.

The current law of the Great Lakes assumes the lakes are fully allocated and that there should be no major new diversions. An International Joint Commission (IJC) report characterizes the lakes as a "nonrenewable resource" because less than 1 percent of lakes' waters are renewed annually by precipitation[40] and concludes that "*if all interests in the Basin are considered, there is never a 'surplus of waters in the Great Lakes system.*"[41] The question is whether this assumption can sustain itself if prolonged droughts occur and regional and non-regional users attempt to tap the lakes.

Lake use is controlled by three overlapping legal regimes, state, federal and international. All three regimes can be characterized as immature legal regimes. The reasons are both physical and institutional. The basin is basically a closed, balanced system. There are only five major in- or out-of-basin diversions. Most diversions are non-consumptive and there is one major diversion into the basin: the Long Lac-Ogoki diversion from the James Bay basin into Lake Superior. The Lakes flow very slowly from Superior to the Saint Lawrence River. At the present time, only Ontario's and Superior's levels are regulated by dams and locks.[42] The lack of regulation is a function of the fact that "*for the most part, the Great Lakes act as a natural system and water will flow through the system only as quickly as nature will allow,*"[43] sometimes 12 to 15 years. For this reason, the rights of users and littoral states remain largely inchoate, with the exception of the Chicago diversion. As a matter of United States federal common law, all littoral states have an equal right to a fair share of interstate waters along or within their borders, but these rights must be claimed and confirmed by a judicial proceeding or by Congressional legislation.

3.1. THE FEDERAL INTEREST

As a matter of legal power, the United States federal government has the dominant interest in the allocation and use of the Great Lakes, but, aside from navigation protection, this interest has been delegated to the states. Constitutionally, the federal government could probably do anything it wants with the Great Lakes, from draining them to re-establish a sea in the Great Basin in Idaho, Nevada, and Utah to dedicating their use exclusively to be Great Basin

States, the rights of the federal government of Canada, and of the provincial governments of Ontario and Quebec aside. The real issues are not, however, on what Congress could do, but what it has done and is likely to do. Federal power over the Great Lakes has followed the pattern of federal power over water resources established in the 19th century. Aside from navigation protection, the federal government has deferred to the state water policy. On the Great Lakes, Congress has allowed the littoral states to develop an anti out-of-basin diversion strategy and ratified it by legislation,[44] which allows states to prohibit new out-of-basin diversions.

3.2. THE STATE INTEREST

By virtue of their ownership of the Lake beds (lands underlying the mean high water mark) and their control of littoral access, the seven Great Lakes states and the Province of Ontario have the primary interest in regulating the Lakes. State and provincial power is, of course, subordinate to the power of the national governments to regulate lake use. The power to regulate lake use is plenary in the United States, but is more circumscribed in Canada due to the greater constitutional powers of the provinces *vis a vis* the federal government. The states have used their political power to control the use of the Lakes in two related ways to protect the integrity of the Great Lakes' aquatic ecosystem and to preserve lake waters for the exclusive use of the basin states. In 1985, they agreed to the Great Lakes charter which requires that all states consult with each other the Province of Ontario before they approve an out of basin diversion under state law.[45] The Charter is a presumptive Congressional exemption of gubernatorial vetoes which prohibit out-of-state diversions from the negative or dormant commerce clause, but its constitutionality has never been tested.[46]

There have been several diversion proposals since 1986 but the Charter has prevented all but small trans-water municipal diversions. The potential use of the Charter to control lake use by preventing out of basin diversions to alleviate a prolonged drought is illustrated by the fate of then Illinois Governor James Thompson's proposal to triple Lake Michigan diversions in the drought summer of 1988. As the Mississippi dropped, barge navigation was impeded, and Governor Thompson wanted the trans-basin diversion to augment the River's record low flow. The proposal, allegedly to help downstate grain exporters who were major campaign supporters, was blocked by protests from Ohio, Wisconsin, Minnesota and Canada, and then Governor James Thompson dropped it in the face of intense interstate and foreign opposition.[47] The chief legal basis of the objections to this proposed quick navigation fix was Illinois' failure to follow the Great Lakes Charter consultation procedures.[48] The presumption that withdrawals are harmful has strengthened over the years. In the late 1990s, proposals to export bulk water from Canada produced a Canadian ban on such transfers.[49]

4. The Florida Everglades: Adaptive Management in Practice

Many river basins have been modified by past dam and diversion schemes at the cost of severe environmental damage. Increasingly, nations are trying to restore these degraded aquatic ecosystems by recreating flow regimes that provide some of the ecological functions and services lost by diversions and dams.[50] These efforts are integral components of IWRM but may adversely effect existing entitlements depending on the water year and physical options. An ambitious joint federal-state effort is underway to restore the former sheet flows to the Florida Everglades National Park which occupies the southern end of the United States federal state of Florida.[51] Restoration will include the construction of new surface water reservoirs, the restoration of wetlands, the use of underground storage to create dry period reserves for urban areas, and the removal of artificial barriers to permit sheetflows of water into the Everglades. This restoration experiment is fraught with uncertainties. For example, many scientists warn that high flows will compromise the ecologically important tree islands in the Central Everglades above the Park[52] and may increase the salinity of Central Florida Bay.[53]

To deal with the inevitable scientific uncertainties and risks to existing entitlements, the Master Plan formally requires the practice of adaptive management.[54] Adaptive management was developed in the late 1970s as a remedy for static or deterministic environmental impact assessment and is increasingly becoming an accepted modern water management tool. The basic criticism of traditional assessment that a fixed review of an independently designed policy[55] is inconsistent with the experience of resource managers worldwide and with what has come to be called non-equilibrium ecology which posits that ecosystems are dynamic not inherently stabile systems. The need for rigorous but flexible procedures to make decisions under conditions of uncertainty has a long intellectual pedigree including decision theory. Many resource decisions always had an experimental component and including monitoring procedures and IWRM can build on this legacy.

Adaptive management can also be understood as an adaptation and modification of the precautionary approach to resource management and environmental protection. AM acknowledges the need to act in face of scientific uncertainty, but it also realizes that the uncertainty must be narrowed over time and that management decisions can no longer be presumed to be final. Management institutions must have the capacity to modify earlier actions as new information about environmental and other impacts is collected and understood. Basically, adaptive management makes the experimental nature of a decision and the need for information collection the primary components of management. Thus, it requires a continuous process of acquiring evaluating and adjusting to new scientific information.

AM also requires the practice of regulatory science. Regulatory science is science designed to answer, to the best extent possible, causal questions about management choices and a socially desired outcome such as the preservation of a species from extinction or an ecosystem that functions more like it did prior human intervention. Once baselines and targets are set, scientists must assess whether they are being maintained. This usually requires a high level of cross-disciplinary integration and informed speculation. It is not enough to collect data. Scientists must be permitted to draw inferences from incomplete data and take appropriate regulatory actions based on these inferences provided an adequate midcourse protection plan is in place. The hope is that AM will minimize the adverse impacts on existing entitlements at the same time that the restoration goals are achieved. Thus, AM should be an integral part of any IWRM exercise because it is a promising mechanism to change aspects of an existing allocation regime in a way that is both scientifically credible and fair.

IV. WATER LAW

IWRM must be supported by effective national and international water rights regimes. The best guarantee that water will be used in an environmentally sustainable and equitable manner to serve the full range of uses from basic human consumption to aquatic ecosystem conservation is an effective national water law regime. In addition to creating relatively secure entitlements, national legal regimes should include provisions that deal with the following problems: (1) the allocation or reallocation of water for the maintenance of aquatic ecosystems services and the restoration of degraded riverine environments, (2) the reallocation of water from marginal agriculture to more efficient uses, both urban and environmental, (3) the creation of innovative reallocation techniques such as water banks, (4) the protection of rural, generally poor, areas that may face the loss of water and livelihood opportunities from large-scale, trans-basin diversions, (5) the protection of minority groups such as indigenous peoples and other who have developed sustainable customary use practices from undue disruption of these uses, (6) the limitation of the mining of groundwater aquifers, (7) the provision of essential water supplies in times of scarcity for a wide range of uses at a time when there is less support for large-scale subsidized supply augmentation, dams, (8) the integration of water quality and quality regulation so that there are adequate flows to complement pollution control strategies, (9) mechanisms to incorporate possible strategies to adapt to global climate change which threatens to alter rainfall patterns and create more extreme cycles of flood and drought, and (10) the development of more adaptive, transparent and inclusive decision-making processes.[56]

The rest of this section includes a selective discussion of some of the most important challenges that nations face in aligning national water law regimes with the teachings of IWRM. The primary focus is on the need to limit unsustainable water uses.

1. Limitations on Consumptive Use and the Redefinition of Entitlements

The pressures for aquatic ecosystem conservation or restoration and the risk of supply and demand imbalances often require that consumptive uses be capped or even rolled-back. South Africa's reserved environmental flows (as well as for basic human needs) and the cap imposed under the Murray-Darling management regime in Australia are two outstanding examples of this trend. These new flow maintenance initiatives have been done within the framework of existing entitlement regimes, but they have the potential to modify and thus redefine them.

Any limitation regime adds a new element of incompleteness to water rights, this should be explicitly recognized. Water rights have always been subject to the laws of nature, the fixed risks of established rules such as priority and the correlative rights of other users, which run from a complete loss through capture to post-use sharing duties, and the high degree of state power, in any legal system, to control the use of water. New demands on the system must be recognized as another risk entitlement holders must anticipate. The inherent incomplete or risky nature of property rights means that the focus should be on the actual expectations that lie behind a use, rather than the perpetual enforcement of the entitlement. This permits regulators, users and other shareholders to explore alternative ways of satisfying those expectations in ways that accommodate new uses can be found.[57] In some cases, IWRM may require the substitution of risk-based physical solution that provides an adequate margin of safety for an absolute or firm entitlement in water-short years. These new regimes will be characterized by the greater reliance on physical solutions,[58] which include adaptive management, and water markets rather than the anticipated enforcement of priorities and formal entitlements.

2. Increased Alienability

Water entitlements have often been the bi-products as ambitious irrigation developments and thus have been viewed as tied to a specific parcel of land or a water resources project, but they are increasingly being made more alienable to correct prior misallocations, primarily the dedication of too much water to low value agricultural use. Alienation potentially makes water rights marketable commodities, but it is essential to recognize that water markets are not an end in-and-of-themselves but a means to the more efficient and sustainable use of water through fair reallocation procedures. The state should decide the amount of water which is subject to reallocation, and markets must be closely monitored to ensure that the transfer of water is not unduly disruptive of local economies and ecosystems and that the transfer result in the actual application of water to a productive use. The Chilean experience with water markets is instructive. Water marketing was embraced as part of the government's enthusiasm for a full market economy. However, studies of the operation of water markets demonstrate that with few local exceptions, market transactions of water rights in Chile have been limited... Up to 1996 no more than 5% of the water rights in highly utilized rivers has been transferred. And of this 90% of the transactions correspond to non-utilized rights, rather than reallocations from one productive use to another.[59]

3. New Ground Water Conservation Regimes

Groundwater conservation is a major problem in many areas because the resource is stressed by over pumping. However, groundwater is more difficult to regulate for three primary reasons. First, the articulated conservation standard, safe yield of a basin or aquifer, is not a simple scientific standard but requires complex decisions about the long term water budget of the system and length of time that an area should use the resource. Second, the adverse impacts of pumping on aquifers and related surface streams materialize over long time horizons compared to the adverse impacts of many surface withdrawals. Third, and related, it is more difficult to incorporate use limitations into groundwater rights compared to surface rights because all pumpers contribute simultaneously to overdraft problems.

The challenges for regulators are to (1) to assemble the necessary information to understand the total hydrologic impacts of groundwater pumping, (2) to integrate ground and surface rights, (3) to limit the excessive mining (extraction in excess of an agreed upon recharge rate) of aquifers which may require that some basins be closed to new wells, and (4) to insure that groundwater pumping does not impair the quality of the aquifer. This is a particular problem in coastal or ancient inland sea areas where pumping may create a cone of depression which causes salt water intrusion into an aquifer.[60] At a minimum, states need the authority to define the sustainable yield of stressed basins, to limit unsustainable withdrawals and to coordinate ground and surface uses.[61]

4. The Integration of Water Quantity and Quality

The regulation of water quality has traditionally been considered a separate activity from water allocation.[62] Water quality regulation limits what can be put in and water allocation law limits what can be taken out of a stream. Of course, the two are connected. Justice Sandra Day O=Connor of the United States Supreme Court has characterized the distinction between quantity and quality as artificial.[63] Water quality is usually measured by compliance with water quality standards. Technology-forcing regulations are widely used to ensure compliance with these standards. The maintenance of water quality standards assumes some minimum flow levels, and thus that withdrawals can compromise water quality.[64] The logic of the connection is clear but few courts have held that new withdrawals must be measured by their water quality as quantity impacts.

V. CONCLUSION

Many nations are modifying their traditional water allocation and planning regimes to adopt elements of IWRM. The diverse, uncoordinated efforts collectively strengthen the argument that IWRM is an emerging principle of international environmental and water law. IWRM is both a process for making better allocation decisions and set of standards against which the decisions can be measured. It is not a substitute for making hard political choices about alternative water use options; it is a focused search for more national and transnational environmentally sustainable, socially equitable and fair alternatives.

Notes

[1] E.g., Report of the World Commission on Dams, DAMS AND DEVELOPMENT: A NEW FRAMEWORK FOR DECISION-MAKING (2000).

[2] Professor Ludwik A. Teclaff was one leading advocates of the need to recognize the benefits of historic flood patterns as well as the benefits of flood control. See Ludwik A. Teclaff, *Treaty Practice Related to Transboundary Flooding*, 31 Natural Resources J. 109 (1991) and The River Basin in History and Law (1967). For a recent analysis of the need to restore the flood functions of a large river see National Research Council, THE MISSOURI RIVER ECOSYSTEM: EXPLORING THE PROSPECTS FOR RECOVERY (2002).

[3] See Daniel Botkin, Discordant Harmonies (1991) and William Jordan, The Sunflower Forest: Ecological Restoration and the New Communion with Nature (2003).

[4] *See generally* Reed E. Noss and Allen Cooperider, Saving Nature's Legacy: Protecting and Restoring Biodiversity (1994). The changes build on the substitution of a non-equilibrium

for an equilibrium paradigm in ecology. See A. Dan Tarlock, *The Non-equilibrium Paradigm in Ecology and the Partial Unraveling of Environmental Law*, 27 Loyola of Los Angeles L. Rev. 1121 (1994) and Fred Bosselman and A. Dan Tarlock, *The Influence of Ecological Science on American Law: An Introduction*, 69 Chicago-Kent L. Rev 847 (1994).

[5] See Notes ___ to ___.

[6] Water in the West: Challenge for the Next Century 3-2 (1998).

[7] *Ibid.*

[8] Jack Stanford, et al., A *General Protocol for the Restoration of Regulated Rivers*, 12 Regulated Rivers: Research and Management 391 (1996).

[9] Peter Gleick has proposed a similar list for global water sustainability. Sustainable water management must include (1) a human right to the minimum amount of water to sustain human health, (2) the recognition of the need for water to maintain and restore ecosystems, (3) the decreased reliance on structural solution such as supply augmentation, (4) the application of efficiency principles to water use, (5) the more efficient design of new water supply and distribution systems, and (6) increased NGO and stakeholder participation in decision making. Peter H. Gleick, *The Changing Water Paradigm: A Look at Twenty-First Century Water Resources Development*, in 25 WATER INTERNATIONAL 127, 131 (2000).

[10] See, e.g., Sarah F. Bates, David H. Getches, Lawrence J. MacDonnell and Charles F. Wilkinson, Searching Out the Headwaters: Change and Rediscovery in Western Water Policy (1993). The current focus is on restoration because most large river systems in the world have been substantially modified. We cannot fully restore degraded systems; the best we can do is recognize that rivers perform valuable functions from the maintenance of consumptive uses to ecosystem services. See Nature's Services: Societal Dependence on Natural Ecosystems (Gretchen Daley ed. 1997).

[11] See Ola Busari, Integrated Water Resources Management in South Africa: Spreading Both Wings of the Capacity-Building Eagle, WARFSA/Waternet Symposium: Integrated Water Resources Management: Theory, Practice Cases, Capetown, SA, October 30–31, 2001.

[12] Ringius, L. et al., *Climate Change in Africa—Issues and Challenges in Agriculture and Water for Sustainable Development*, CICERO Report 1996:8.

[13] Dams and Development, *supra* Note ___ at 68.

[14] Dams and Development, *supra* Note ___ at 77–78.

[15] See National Research Council, 1999. WATER FOR THE FUTURE: THE WEST BANK AND GAZA STRIP, ISRAEL AND JORDAN.

[16] See THE MISSOURI RIVER ECOSYSTEM, *supra* Note 2.

[17] Report of the Expert Group Meeting on Strategic Approaches to Freshwater, Para 11 (1998) <http://www.un.org/documents/ecosoc/cn17/1998/background/ecn171998Bfreshrep.htm> (visited 12-1-02).

[18] International Year of FreshwaterBUN General Assembly Resolution A/RES/55/196, 1 Feb 2001, <http://www.unesco.org/water/iyfw/res/_55_196e.pdf> (visited 12-1-02).

[19] Ministerial Declaration adopted my ministers meeting in the Ministerial Session of the International Conference on Freshwater Bonn, 4 December 2001, <http://www.water-2001.de/outcome/Ministerial_declaration.asp> (visited 12-1-02).

[20] Johannesburg Declaration on Sustainable Development (2002) Art. 5.

[21] Agenda 21, Chapter 18, Para 18.6. In many countries, fragmented and incomplete water authority frustrates these objectives. See S.M.K. Donkor and Yilma E. Wolde, INTEGRATED WATER RESOURCES MANAGEMENT IN AFRICA: ISSUES AND OPTIONS (United Nations Economic Commission for Africa).

[22] Raymundo Garrido, *Water Resources National Policy in Brazil*, paper prepared for World Commission on Dams. Http://www.dams.org.

[23] Patricia Wouters et al., *The New Development of Water Law in China,* ___ Denver Water L. Rev. ___ (2004).

[24] EU Water Framework Directive 2000/60/EC.

[25] Act 36 of 1998.

[26] The leading case is National Audubon Society v. Superior Court, 658 Pacific Reporter 2d 709 (Cal. 1983).

[27] Section 1, Act 36 of 1998.

[28] Michael Kidd, South Africa's National Water Act: A Five-Year Report Card, in 1 Law, Water, and the Web of Life 177, 187 (Institute of Law for a Green Planet Antonio Herman Benjamin ed. 2003).

[29] See A*General Protocol for the Restoration of Regulated Rivers*, supra note 8.

[30] The Murray-Darling Basin Agreement, available at www.mdbc.gov.au.

[31] Murray-Darling Basin Agreement, Part X.

[32] See Poh-Ling Tan, *Irrigators Come First: Conversion of Existing Allocations to Bulk Entitlements in the Goulburn and Murray Catchments, Victoria*, 18 Environmental and Planning Law Journal 18 (2001).

[33] 1998–1999 Annual Report, *supra* Note ___ at 24.

[34] Murray-Darling Basin Commission, Murray-Darling Basin Cap on Diversions: Water Year 1997/98 Striking the Balance (December, 1998).

[35] Water Audit Monitoring Report 1996/97, Report of the Murray-Darling Commission on the Final Year of the Interim Cap in the Murray-Darling Basin (October, 1998).

[36] Water diversions in the Murrumbidgee Valley are approaching the upper confidence levels of the CAP irrigation is projected to increase. Review of CAP implementation 1997/98 Report of the Independent Audit Group (November, 1998).

[37] In 1999, the Murray-Darling Ministerial Council commissioned a comprehensive review of the CAP. Http://www.mdbc.gov.au/naturalreso . . . strategies/projectscreen/the_cap.htm. See Notes ___ to ___, infra.

[38] 8 Inland Rivers Network News, No. 3, Summer 2003.

[39] The procedure is outlined The Pilot Interstate Water Trading Project, http://www.mdbc.gov.au/naturalreso . . . rojectscreens/pilot_watertrade.htm

[40] PROTECTION OF THE WATERS OF THE GREAT LAKES at 4.

[41] Id. at 26.

[42] Michael J. Donahue, Alicia A. Bixby and David Siebert, *Great Lakes Diversion and Consumptive Use: The Issue in Perspective*, 18 CASE W. RES. J. INT'L L. 19, 25-26 (1986).

[43] GREAT LAKES COMMISSION, WATER LEVEL CHANGES: FACTORS INFLUENCING THE GREAT LAKES 5 (1986).

[44] 43 U.S.C.' 1926d-20 requires the consent of all of the littoral state governors to an out-of-basin diversion.

[45] *See* Peter V. MacAvoy, *The Great Lakes Charter: Toward A Basinwide Strategy for Managing the Great Lakes*, 18 CASE W. RES. J. INT'L L. 49 (1986).

[46] A widely circulated joint Canada-United States legal study prepared for the Great Lakes Governors has concluded that the Water Resources development Act of 1986 violates, inter alia the dormant commerce clause, the non-delegation doctrine and the due process clause. See Joseph W. Dellapenna, *The International Joint Commission Considers Water Exports From the Great Lakes*, 3 ABA Water Resources Committee News Letter, No. 2, page 10, January, 2000.

[47] Stanley Chagnon & T. Karl, Drought and Natural Resources Management in the United States: Impacts and Implications of the 1987–89 Drought 43–112 (1991).

[48] 28 Canadian Yearbook of International Law 407-409 (1989) contains a summary of the Canadian parliamentary debates in opposition to the proposal.

[49] An Act to Amend the International Boundary Waters Treaty Act, S.C. 2001, c.40.

[50] I have addressed the problems in defining restoration objectives in A. Dan Tarlock, *Slouching Toward Eden: The Eco-Pragmatic Challenges of Ecosystem Revival*, 87 Minn. L. Rev. 1173 (2003). See generally Eric Higgs, Nature by Design: People, Natural Process, and Ecological Restoration (Cambridge MIT Press, 2003).

[51] U.S. Army Corps of Engineers and the South Florida Water Management District, COMPREHENSIVE EVERGLADES RESTORATION PLAN: MASTER PROGRAM MANAGEMENT PLAN (2000).

[52] Keith Kloor, *Everglades Restoration Hits Rough Waters*, 288 Science 1166, May 19, 2000.

[53] National Research Council, FLORIDA BAY RESEARCH PROGRAMS AND THEIR RELATIONSHIP TO THE COMPREHENSIVE EVERGLADES RESTORATION PLAN (2002).

[54] Master Plan at 8.

[55] Holling, C.S., ed. 1978, Adaptive Environmental Assessment and Management, p. 136.

[56] See Burchi, *Water* 150–159 for a survey of recent legislation dealing with these issues.

[57] This is consistent with the United States Supreme Court's taking jurisprudence. E,g, Pallazolo v. Rhode Island, S33 U.S. 606 (2001).

[58] Technically, physical solution exchanges wet water for the right to assert water rights that would promote inefficiency. See Harrison Dunning,

[59] Miguel Solanes, *Water: Rights, Flexibility and Governance: A Balance That Matters?* (April, 2003). See generally Carl J. Bauer, Siren Song: Chilean Water Law as a Model for International Reform (2004).

[60] See Tarlock et al., WATER RESOURCE MANAGEMENT, *supra* Note—at 532–546.

[61] E.g., North Carolina Statutes Annotated S 143-215.13.

[62] See David H. Getches, Lawrence J. MacDonnell & Tereas A. Rice, Controlling Water Use: The Unfinished Business of Water Quality Protection (1991). For an analysis of the integration of water quantity and quality in South Africa see Andrew Allan, A Comparison Between Water Law Reforms in South Africa and Scotland: Can a Generic National Water Law Model Be Developed From These Examples?, 43 Nat. Res. J. 419, 442–446 (2003).

[63] The United States Supreme Court characterized the distinction between quality and quantity as artificial. PUD No. 1 of Jefferson County v. Washington Department of Ecology, 511 U.S. 700 (1994).

[64] The Clean Water Act does not allow dischargers to obtain a credit for discharging into clean water. Weyerhaeuser Co. v. Costle, 590 F.2d 1011 (D.C. Cir. 1978).

INTEGRATED WATER RESOURCES MANAGEMENT: EXPERIENCE IN THE ARAL SEA BASIN

CHARACTERISTIC FEATURES OF INTEGRATED WATER RESOURCES MANAGEMENT IN THE SYRDARYA RIVER BASIN

M. Kh. Khamidov
Director of BVO "Syrdarya," Republic of Kazakhstan

Water allocation management in areas of such critical water deficit as the Aral Sea basin has always presented many complex problems. It requires the coordination of the water demands of all users within the basin with the restrictions imposed by the water supply regime. The water users in the basin include irrigated agriculture, industry, drinking water suppliers, hydropower, fish husbandry, ecosystem protection, recreation and flood control measures.

The process of effective water management is particularly complicated in the Syrdarya basin because of several unique conditions:

- even in years of high water flow, water resources are completely exhausted, without any reserve;
- the degree of complete, long-term river flow regulation is very high: $K = 0.93$;
- the amount of water used exceeds the available basin river flows and is covered only through the reuse of drainage water from irrigated fields.

These factors have combined to cause serious problems with respect to river water quality, in particular in relation to the growth in. In addition, strict limits have been imposed on water allocation in the four riparian states. In recent years the transition of the Toktogul reservoir design regime from irrigation power to power alone, combined with a corresponding sharp increase in power releases in the fall—winter period, have aggravated the already difficult problems of Syrdarya water resources management.

It is well known that the most comprehensive and expedient way of ensuring that river flows meet the demands for water consumption is through regulation. For this purpose, reservoirs are constructed on the river to distribute water in the amounts and at the times needed by water users. Consequently, water discharges are more precise, with less waste, than under the natural regime.

The reservoir at Toktogul is the key element in the Syrdarya flow regulation, and Naryn–Syrdarya cascade regulation provides compensating, long-term regulation, although only if the design regime of its operation is strictly adhered to.

P. Wouters et al. (eds.), Implementing Integrated Water Resources Management in Central Asia, 25–34.

Water use in Syrdarya basin assumes that the operation of the main regulator—Toktogul reservoir—is based on supplying the amounts of water guaranteed to the water users in each republic, allowing sustainable agricultural crop yields. Therefore, three quarters of the annual amount of water released from the Toktogul reservoir is carried out in the growing period (9.43 km^3). Following the general water use irrigation regime, Toktogul reservoir should release less than 180 $m^3\ s^{-1}$ (2.85 km^3) in the growing period, which corresponds to hydropower generation's minimal capacity.

The transition to a power-focused regime at the Toktogul Waterworks has changed the water availability situation in Syrdarya basin. As a result, the maximum generation of the hydropower plant now corresponds to the winter period, and in the growing period releases from the reservoir are reduced to allow the accumulation of water. Similar modification of the operation schedule has led to a comprehensive deformation of the river regime.

The following negative effects of the Toktogul reservoir operation regime can be noted:

- acute water shortage for irrigation in growing period;
- irreversible river water losses occur and increase the need for releases in the Arnasay sink, as a result of having no ability to pass winter releases to the Aral Sea because of the freezing situation in the Syrdarya's lower reaches. Releases in Arnasay damage the flooded area and its infrastructure;
- ecological sustainability of Syrdarya aquatic systems has been lost—in winter the river channel is flooded and is not able to divert return water. In summer a critical ecological—epidemiological situation emerges, as the river channel, in particular in its middle reaches, is close to desiccation;
- with the number of dry years set to increase under the new operation regime as the Toktogul reservoir is emptied, it will not be possible to regulate Syrdarya flow over the long term. Inevitably, electric power production will decrease simultaneously.

The normal functioning of the water management system in the Syrdarya basin has failed, and its capacity and workability has been reduced. In recent years, basin management and water use has been complicated by a number of factors, the most important of these being:

- continuous population growth in the basin (by between 2.2 and 2.5% per year), effectively doubling by 2020, with a commensurate increase in water demand. Consequently, demographic conditions are determinative;
- protection of ecological requirements caused by the Aral Sea problem and aggravation of the general situation in Central Asia, including water quality deterioration, that in turn leads to a lowering of the level of water that is available for use;

- after 1991, engineering structures at the interstate and national levels were owned by different sectors and owners, resulting in contradictory measures being taken, and the inevitable water losses became more frequent.

The above-mentioned factors may result in the water resources available per capita being more than halved. Without a considered combination of technical and organizational measures, the ability of the Central-Asian population to survive must be in question.

In these conditions, trans-boundary water use can be very significant. It should be rational, effective, and equitable. Water resources management should be sustainable and based upon ecosystem requirements. It should also meet the needs of the current population without damaging other states and without jeopardizing the needs of future generations. Sustainable development assumes the improvement of society's wellbeing without negative environmental impacts, along with the achievement of a constant and continuous balance between uses and resources. Therefore, the connection of these two factors provides the basis for an analysis of water resources formation on the river trunk, amplifications of water demands, and the economic situation in the region.

The principal problem in water resources planning and management is balancing the water available from the Naryn–Syrdarya with the demands made on the river by all the basin countries. At the same time, the most effective method of developing and improving the water management system is the transition to integrated water resources management (IWRM). Integration of all hierarchical water management levels (BVO, Ministry of Water Resources, hydro-meteorological services, water and canal departments, WUA, farmers and others) becomes especially important in the context of IWRM.

To solve this problem it is necessary to define priority actions and envisage new management mechanisms:

- organizational, legal, and regulatory mechanisms must meet the requirements of all regional states. It must be noted that this approach, and the IWRM method which stimulates it, seeks to fulfil functional responsibilities: for example, water supply activity must meet the need for economic development;
- factors such as landscape conservation and the prevention of emergency events demand the establishment of a set of interacting institutions with specific powers that can stimulate water management activity. This approach is therefore appropriate for implementing long-term programmes that facilitate the elimination of the contradictions between the natural variations in river water availability and agriculture needs. By creating an opportunity to use water resources more widely and more effectively, river flow regulation

promotes the resolution of two main problems—firstly, by increasing river water availability in low water periods, and secondly, by lowering flood elevations. This means that water users are guaranteed more certain and higher rights of use than under the natural regime, with minimal water discharge.

In general, IWRM is a process which can provide coordinated development and management of water, land, and other associated resources, in order to maximize economic and social wellbeing based on principles of equity, damage reduction, and ecosystem viability.

It should be recognized that IWRM does not provide a single interstate water management framework, but rather a holistic approach to the improvement of water management structures integrated with the maintenance and development of national management systems. It takes account of the requirements not only of the water regime, but also of the strict limitations on the discharge of pollutants.

In order for IWRM to be realized in practice, some time will be required to allow certain organizational issues to be overcome. These include:

- national and administrative boundaries and constraints;
- sectoral restraint;
- gap between hierarchical water management levels;
- absence of mutual interests on part of, on the one hand, water users and, on the other hand, the water regulating bodies;
- lack of interest of the managers of water management structures in the productivity of water resources;
- poor personnel provision and training of experts;
- low technical level of structures and equipment;
- unstable financial support of operation and development.

At first glance the transition to IWRM seems to be relatively simple. In practice however, it is a long and complex process which demands the radical restructuring of existing organizational structures of the water hierarchy of each state. At the same time, it is also necessary to take into account specific features of the water infrastructure as well as the specific importance of irrigated lands and the amount of water used in each state.

Improvement of water resources management must integrate the following set of measures:

- Development of mutually beneficial agreements. These should address the water intake limits of the republics; the Naryn–Syrdarya reservoirs cascade operation regime; and the long-term regulation of the Toktogul reservoir.
- Stage-by-stage implementation of a water resources-use mechanism, based on cost recovery.

- Water resources account improvement, based on the dissemination and use of modern advanced technologies, allowing for the automation and remote control of hydro-structures and canals.
- Maintenance of engineering structures and main canals as well as all BVO "Syrdarya" infrastructures, in a technically usable condition.
- Increase in investment in the maintenance, development, and improvement of water management systems.
- Transfer of all water intake structures along the Syrdarya trunk to BVO "Syrdarya" for temporary use.
- Provision of a solid legal basis for the BVO.

Recent interstate agreements governing water resources use, concluded only with respect to the growing period, are unable to solve the long-term problems of Totktogul reservoir regulation and effective Syrdarya water use. In recent years, the Toktogul reservoir operation regime has been carried out under conditions of severe low water availability. Reservoir inflow by the beginning of growing period often does not meet with the Glavgidromet forecast. Water inflow sometimes exceeds the norm, and Toktogul reservoir cannot be released up to dead volume. It is therefore necessary to carry out long-term Totktogul reservoir regulation such that releases in the winter period are reduced to 5.5 km^3, with proper compensation supplies being available to Kyrgyzstan to enable it to implement its irrigation-power regime.

One of the problems has been that a proper compromise has not been achieved between republics on common or differentiated tariffs for energy resources. Determination of the cost of 1 kW of electric power, and agreement on the principles of long-term Toktogul reservoir regulation, would be of benefit to all Syrdarya basin states.

In the last year, different recommendations have been suggested on interstate agreements and energy resources tariffs at world prices between Uzbekistan, Kazakhstan, and Kyrgyzstan. However, not all of these agreements have been signed.

BVO "Syrdarya" proposes an absolutely new approach and mechanism of interstate cooperation on integrated water resources use, which presents transparent and confidential documentation. The Republics of Kyrgyzstan, Kazakhstan, Tajikistan, and Uzbekistan, with the involvement of their own water and power design institutes, would independently develop feasibility studies (FS) of water resources use, which should be based on long-term regulation (10–12 years) of Toktogul reservoir water amounts, integrating with economical indicators of their own country, produced electric energy, compensation supplies, and water intake limits.

After these feasibility studies are completed, design institutes would exchange their projects, bring appropriate amendments in FS, and after coordination submit them for working group consideration, which would consist of representatives from all republics. The working group, after consideration and completion, would present this document to the Governments of the Republics, allowing them to review the long-term interstate agreement. The development of feasibility studies, compensation supplies, and pricing should be based on compromise between the nations, without the need to involve international or outside mediators.

Another question surrounds direct supplies of thermal and power resources as compensation for non-produced energy. These should be made by the organization, which could attract the states. It is high time to create a waterpower consortium, which has already been approved. This structure should include BVO "Syrdarya" and UDC "Energiya" as regional units responsible for basin management by waterpower resources use.

Analysis of the Naryn–Syrdarya cascade indicates the necessity of improving operative water management as well as increasing water account correctness and river water quality through the use of advanced world technologies. Water accounting is carried out on the Rivers Naryn, Karadarya, Syrdarya, Chirchik, and Isfara, in 430 water intakes, including 187 intakes under the charge of BVO, making BVO "Syrdarya" the experts responsible for water supply to Syrdarya basin republics. During the vegetation period, at all major water intake structures except BVO subdivisions' experts, workers of BVO central unit permanently monitor the situation to avoid stress on linear organization workers by local bodies.

To implement operative management, waterworks should be provided with control-metering devices and communication systems corresponding to present levels. These allow:

- Decisions to be made regarding certain tasks in waterworks regulation, including automatic maintenance of established regulation parameters;
- data collection on controlled parameters and data reliability control;
- increase in effective control and management;
- test control of all technical means of complex.

All of this will allow implementation of remote control measurement of water levels, discharges, and electric conductivity, as well as the opening of structure gates, continuous collection, storage, and processing of remote control data, automated regulation of water levels and discharges, the remote control of specific gates, maintenance of continuous communication with dispatch stations, and liquidation of equipment failures.

Taking into account all of the above mentioned, we can see that the dissemination of advanced technologies on headwork structures charged to BVO "Syrdarya" will increase water management efficiency, including control of water losses, as well as land and water productivity. A great deal has already been done in this respect. For example, with technical assistance provided by foreign donors, BVO "Syrdarya" has introduced advanced technologies on a set of objects.

The Kyrgyz company "Sygma" fulfilled similar works at Uchkurgan Waterworks, Naryn River, whilst at Verkhnechirchik Waterworks, Chirchik River, this development was undertaken with the financial support of the American Agency USAID and the Swiss International Agency. Similar work has taken place at South Golodnostep Canal headwork under EC IFAS support. The radio-telemetering system "SCADA" was installed, which allows the user to obtain information on water discharges up to 50 km from the object in real time, and also control gates on "Dustlik" Canal headwork (former Kirov's Canal). The Canadian Engineering Company UMA-international produced and implemented this system. Putting automated water management systems into operation in Syrdarya basin has demonstrated the potential and reliability of this technology, and has promoted water-metering correctness both in gauging stations' data and algorithms of gate opening control and water heads control. In the future, it is planned to automate Kujganyar Waterworks, as well as structures on the "Khakulabad" and Big Andizhan Canals.

Regarding the preparation and training of BVO "Syrdarya" staff, in 2002, under USAID aegis 64, BVO experts were trained in the new technology. This year, a further fifty experts were trained in optimal computer-modeling of water problems solutions.

The introduction of automated and operative water sector control systems has had a further benefit, in that the putting into practice of these systems has resulted in an increased confidence between Central-Asian states.

It is also important to examine the current state of canals and hydraulic structures. In the period 2002–2003, examinations of the technical state, maintenance of equipment, and current condition of engineering structures were carried out, in order to provide a thorough overview of operational reliability and safety. Mechanized cleaning was implemented on BFC, KDP, and Dustlik Canals, whilst immediate repairs were carried out on the headwork of Bekabad, VDC, NDC and KDP Canals, and Kujganyar Waterworks. Total excavation work amounted to 297 th. m^3 of concrete and reinforced concrete work, and 353 m^3, 22 tons of metal structures were installed. The communications system was reconstructed on structures of the Golodnostep Administration, where a new re-transmitter, nine permanent and five mobile radio stations were installed. Automation projects have been planned at Kujganyar Waterworks, with work scheduled to start in October 2003. However, the financing of BVO "Syrdarya" water system maintenance is clearly insufficient, and there are practically no allocations of funds for construction and reconstruction of the water services infrastructure.

Presently, depreciation of BVO basic assets and its infrastructure is above 60%, and 50% of the main infrastructure has served out its standard period, with one such example being the 44 km stretch of the Big Ferghana Canal head reach. Similarly, Kujganyar Dam on the Karadarya River is in need of reconstruction, whilst additional cascade structuring is required below Verkhnechirchik Waterworks on Chirchik River. Dustlik Canal requires a large amount of cleaning work, as well as some structural repair. However, BFC and Dustlik Canals transport water almost all-year-round, leaving no time to repair the structures and the canals themselves. Within the allocated means, only emergency and prior repair—rehabilitation works are fulfilled, in order to prevent emergency situations.

The main amount of work on the maintenance of canals and structures to keep them in their proper technical state includes both current and capital repair, the financing of which should be increased by a minimum of 2.5–3 times. In 2002, current repairs were fulfilled to the amount of 368.1 million sum, or 4,10,000 US$. For capital construction of structures and canals, no moneys were allocated. For example, due to a lack of finances, last year BVO "Syrdarya" did not purchase water-metering equipment. Plants in Central-Asian states do not use current meters, with work instead being carried out on aged equipment that is already beyond repair should it break down.

BVO financing is fully dependent on national state budgets, which prohibits any long-term planning which would allow for the creation of strategic investment trends. To be successful in IWRM conditions, it is necessary not only to ensure financing in order to maintain a sustainable regime of structures operation, but also to allow BVO to be flexible in financial regulation while emergency repair works are carried out. Financing priorities should be the rehabilitation of main structures, such as water intake, and regulating structures of interstate importance. At the same time, interrelations should be established as before, on the basis of the Almaty agreement 1992, according to which "basin water organizations are financed at the expense of payments of republican water organizations in equitability and share participation aspects." Big investments are required to maintain and improve all water supply systems and elements.

We will now turn to examine the situation specific to the Aral Sea and Priaralie area. Here, main reserves can be used after water conservation programme implementation. First of all, a modification of the management structure itself is demanded. As things stand, the reach from Chardara to the Aral Sea does not fall within the BVO command zone. ICWC decisions, along with water intake limits, apply, and water should be transported by Syrdarya to Aral Sea. However, BVO is unable to fulfil this ICWC task, if it can be implemented downstream, outside the organization's activity zone. As a result, if water intake limits are approved and amplified annually only because of a sufficient change in year water availability, the Aral Sea's share varies widely. This proves, first of all, that the current structure is insufficiently grounded.

The water situation can be aggravated if appropriate measures on BVO activity improvement are not taken. One suggested solution was to create a united BVO by combining BVO "Syrdarya" and BVO "Aral–Syrdarya," which includes Kyzylorda and Chardara administrations in Kazakhstan as structural subdivisions. This united BVO should consist of representatives of all Syrdarya basin republics as deputy technical directors. It is also foreseen that, in the future, some further reorganization of subdivisions will be required.

One serious problem for BVO experts in operative water management and main canals operation is the crossing of state boundaries on the Big Namangan, Levoberezhny Naryn, Big Ferghana, Verkhny Dalverzin, and Dustlik Canals, as well as boundary crossings on the Naryn and Syrdarya Rivers. It is necessary, therefore, to act to simplify the crossing of state boundaries by basin organization experts without the need for visas and normal controls.

It is worth noting the role which BVO played, and its importance in increasing the efficiency of Syrdarya basin trans-boundary water resources management, effective river flow regulation without conflicts, and guaranteed water availability in the republics according to established water use limits.

Presently, in connection with the change and aggravation of the water situation in the basin, as well as changing social and political conditions, it is necessary to strengthen BVO and ICWC status, and to consider them as international organizations under UNO aegis by the signing of an Agreement between the governments of the five Central-Asian states. This should lead to the reinforcement of the organizational structure for management, conservation, and development of trans-boundary water resources in Aral Sea basin.

The above-mentioned BVO structure will allow the management of water resources in a way that is more effective and transparent, and yet which retains a confidential atmosphere. After the collapse of the USSR, and the subsequent formation of new independent states in Central Asia, the water commission was organized under the initiative of the leadership of these countries to coordinate waterpower resources use. ICWC managed to provide a business relationship for water and power bodies of the republic, avoiding the risk of chaos and confusion in water use.

The trans-boundary waterways management system should be raised up to the level of International Waterpower Consortium. To realize this idea, harmonization of legislation and legal acts of CAR countries is required. The necessity of information exchange and the creation of national and regional databases of integrated trans-boundary water resources use and conservation in the Aral Sea basin is one of the most significant conditions of Integrated Water Resources Management.

Another assessment criterion should be related to agriculture, as agricultural production is the most important component of any human life support system. As such, not everything can be measured by means of economy, especially since the majority of Syrdarya basin's population is engaged in agricultural production, and therefore a lack of water for irrigation purposes means a lack of opportunity to survive. Therefore, it is necessary to provide effective Syrdarya water use, to organize supplies and to stabilize the Naryn–Syrdarya reservoir cascade operation, which would ensure the maintenance of normal living conditions for millions of people of the region, as well as leading to the strengthening of normal relations between the region's states.

In conclusion, it is important to note that the participation of all concerned parties in the process of water resources planning, development and management is the main and obligatory condition for any successful implementation of Integrated Water Resources Management.

PARTICULAR CHARACTERISTICS OF INTEGRATED WATER RESOURCES MANAGEMENT (IWRM) IN THE AMUDARYA RIVER BASIN

Yu. Khudaiberganov
Director of BWO "Amudarya," Republic of Kazakhstan"

In this paper I would like to share my vision of the particular characteristics of the system of integrated trans-boundary water resources management in the Amudarya river basin, based on my experiences as Director of BWO "Amudarya.". Water resources in Central Asia are drawn primarily from the basins of the two largest Central Asian rivers, the Amudarya river and the Syrdarya river. These two rivers provide water to the populations of five states: Kazakhstan; Kyrgyzstan; Tajikistan; Turkmenistan; and Uzbekistan. For the purposes of this paper, I wish to concentrate principally on the Amudarya river basin.

The Amudarya river basin covers a broad area of around 1,327 thousand km^2, of which 1,018.6 thousand km^2 are located in Central Asian states (Turkmenistan—488.1 thousand km^2; the Republic of Uzbekistan—388.2 thousand km^2; the Republic of Tajikistan—131 thousand km^2; and the Kyrgyz Republic—11.3 thousand km^2). The rest of the basin's catchment area is located in Afghanistan and Iran. Meanwhile, independent basins which are located inland but gravitate toward to the Amudarya river are formed by the rivers Kashkadarya, Zarafshan, Murgab, and Tedjen, which lose their hydrological links with the main river.

The Amudarya itself is the largest river in Central Asia. Its length is 2,540 km, measured from the Pyandj river head to the Aral Sea. It is formed at the junction of the Pyandj and Vakhsh rivers. Two large right-bank tributaries (the Kafirnigan and the Surkhandarya) and one left-bank tributary (the Kunduz) flow into its upper reaches. Further down stream, towards the Aral Sea, there is no such inflow to the river. The Amudarya is a river with a glacier-snow catchment area. Therefore, maximum flow is found in the summer, whilst the river reaches its minimum flow around January and February. This seasonal variation in the river flow is quite favorable for irrigation purposes. Whilst crossing a plain from Atamurat to Takhiatash, the Amudarya river loses most of its flow through evaporation, seepage, and irrigation. As to the question of water turbidity, the river is first in this respect in Central Asia and, indeed, is amongst the highest ranked worldwide. The Amudarya river flow is formed in Tajikistan, and then follows a course along the border between Afghanistan

P. Wouters et al. (eds.), Implementing Integrated Water Resources Management in Central Asia, 35–43.

and Uzbekistan, crosses Turkmenistan, flows back into Uzbekistan and runs finally into the Aral Sea. The total average annual volume of water resources in the Amudarya river basin amounts to 78.4 km^3, including 62.1 km^3 in the Amudarya river flow itself.

As regards morphological and geographic characteristics, the Amudarya basin is divided into three sections: the upper reaches (from the Kelif hydropost upstream to the border between Turkmenistan and Uzbekistan), the middle reaches (between the Kelif hydropost and Tuyamuyun), and the lower reaches (downstream from Tuyamuyun). The total irrigated area in the basin measures some 4.5 million hectares. Land in Tajikistan, Uzbekistan (the Surkhandarya oblast), and Kyrgyzstan (a small irrigated scheme in the south of the republic) are irrigated in the upstream area. These irrigated schemes are located in such tributary valleys as the Pyandj, Vakhsh, Kafirnigan, Surkhandarya, and Sherabad.

As to the middle reaches, the largest schemes of modern irrigation are concentrated on extended canals such as the Garagum Canal in Turkmenistan, and the Karshi Main Canal (which is served by six pumping stations), and the Amubukhara canal, both of which are located in the Republic of Uzbekistan). The system of each canal includes off-stream reservoirs. The irrigation systems located between the Kelif hydropost and Tuyamuyun receive water transported by a dozen of canals with non dam-water intakes. In the lower reaches downstream, both banks of the Amudarya river have seen large systems of canals constructed. The Left-Bank Big Canal, the Right-Bank Big Canal, the Turkmendarya, the Klychniyazbay, the Daryalyk, the Khan-yab, the Kyzytken, and the Suenli, were all constructed to irrigate lands in Turkmenitan and Uzbekistan. Major water users in the Amudarya river basin are the power sectors, fishery industries, and those using the basin for navigation purposes. The river waters are also used in industrial and public water supply, and for irrigation purposes.

According to a general agreement between Central Asian states, only stem streams of the Pyandj river, the Vakhsh river, the Kafirnigan river and the Amudarya river itself are subject to interstate water management and allocation. This is a feature unique to this area, and is accepted as a basis for the coordinated interstate management of trans-boundary waters at the regional level. The Agreement, signed by all five countries in February 1992, commits to cooperation in the field of joint management, use and conservation of water resources in international sources, as well as encompassing other documents and deeds adopted by Central Asian states regarding specific river basins on the basis of earlier agreed water allocation schemes. These schemes, along with the above mentioned Agreement, serve as a legal basis for joint water management and allocation among water consumers in the Aral Sea basin.

It is worth noting that the current framework of international cooperation in integrated water management in the Amudarya river basin involves the following Central Asian states: the Kyrgyz Republic; the Republic of Tajikistan; Turkmenistan; and the Republic of Uzbekistan. Since the Kyrgyz Republic consumes less water (450 million m^3), the major water consumers are the Republic of Tajikistan (9.5 km^3), Turkmenistan (22.0 km^3) and the Republic of Uzbekistan (23.2 km^3).

Regarding the Aral Sea and river basins, the organizational framework for interaction between the interstate water management bodies and with national agencies are coordinated with the IFAS framework through the ICWC and its executive bodies (such as BWOs and the SIC). These provide the main channels of interstate cooperation. This is the second unique feature of water management in the Amudarya river basin, since there is no other such organizational framework in the world. Interstate regulation in the Amudarya basin is principally based on the allocation of water limits. This is one of fundamental characteristics of the interstate cooperation.

The following organizations cooperate in water management in the Amudarya river basin:

1. IFAS's organizations:
   - IFAS Board
   - BWO "Amudarya"
   - SIC ICWC
   - AMC ICWC (Advisory Metrological Center of ICWC)
2. United Dispatch Center (UDC) "Energy"
3. National hydro-meteorological services

Existing water systems in the Amudarya basin are characterized by complex structures that provide water conveyance, flow transformation, water intake and supply to consumers, power generation, water quality and quantity control and measurement. All interstate (trans-boundary) water resources, including surface and return waters, fall within the scope of management. Ownership and management of the Aral Sea basin's infrastructure is shared by the national governments and the BWOs. Here, we consider BWO Syrdarya and BWO Amudarya, which are under the supervision of ICWC. It is proposed that the regional infrastructure should include basin infrastructure, which is under the jurisdiction of the BWO. National infrastructure should include all the rest up to farm level. Central Asian states mandated the BWO temporal responsibility for head intake structures located on the Amudarya river and its main tributaries, as well as interstate canals. In addition, the BWO controls all pumping stations, taking water from the rivers and the main interstate canals, in-channel basins, key hydroposts, and the disposal of return waters to the Amudarya, as well as a part of river water intake structures for which BWO bears no responsibility.

As of 1 January, 2003, Basin Water Organization "Amudarya" has borne responsibility for 90 hydrostructures, of which 35 are head intake structures, as well as 337 km of interstate canals, 169 hydroposts, more than 330 km of communication lines, 361 km of inspection roads, 115 km of power lines, 120 radio stations, and 15 radio-relay stations. These represent the bulk of the capital assets used for water regulation and allocation among the states' water consumers.

The objective of BWO is optimal interstate and inter-sector water allocation, in order to meet the water demands of populations and economic sectors, according to water limits approved by the ICWC and on the basis of water availability and the environmental situation. Besides this, the BWO exercises day-to-day monitoring of water withdrawal limits and implements a set of accompanying organizational and technical measures, as well as delivering water to Priaralie and the Aral Sea. To perform the day-to-day trans-boundary water management, four territorial branches of the BWO "Amudarya" have been established in Kurgan-Tyube (Tadjikistan), Turkmenabad (Turkmenistan), Urgench (Uzbekistan), and Takhiatash (Karakalpakstan). These branches have several defined areas of jurisdiction. Verkhnedarya branch operates intake structures and monitors water withdrawals from the rivers Vakhsh, Pyandj, Kafirnigan and from the Amudarya river reach, 246 km long, stretching to the Kelif hydropost. Srednedarya branch monitors withdrawals on the reach, 552 km long, stretching from Kelif to the Darganata hydropost. Amudarya Interstate Canal Authority (Upradik) operates 11 intake structures and 52 hydraulic structures at the main canals, operates and maintains 337 km of main canals, and monitors water withdrawals on the river reach, 167 km long, from Tuyamuyun waterworks to Kipchak hydropost. Upradik also controls and monitors three large irrigation systems, namely the Tashsakin system, the Klichniyazbek system, and the Kipchak–Bozsu system. Nizhnedarya branch operates the Takhiatash waterworks and head intakes to the Khan-yab and Jumabaysaka canals, and monitors all water withdrawals from the river within the reach, 283 km long, from Kipchak hydropost to the Aral Sea. The current water resources operation and control system in BWO Amudarya allows us, to some degree, to cope successfully with the tasks set before the BWO.

It is important to point out that since 1992, under new economic and political conditions, the Interstate Commission for Water Coordination (ICWC) has set water limits for a hydrological year (i.e. for growing and non-growing seasons). States of the region have set the following water withdrawal limits (quotas) for a normal water availability year:

A. Upstream (under the jurisdiction of Verkhnedarya branch)
   1. Kyrgyz Republic—0.45 km^3;
   2. Republic of Tajikistan—9.5 km^3;
   3. Surkhandarya oblast—1.20 km^3;

Total upstream—11.15 km^3.

B. Amudarya midstream and downstream (midstream—under the jurisdiction of Srednedarya branch; downstream—under the jurisdictions of Upradik and Nizhnedarya branch)
   1. Turkmenistan—22.0 km^3;
   2. Uzbekistan—22.0 km^3.
C. Provision of in-stream needs and river water supply to Priarailie
   1. Provision of in-stream needs in the lower reaches during non-growing period—0.8 km^3;
   2. River water supply to Priaralie and the Aral Sea—3.5 km^3;

Total limits for allocation through the Basin during a hydrological year—59.45 km^3.

BWO Amudarya is responsible for the submission of information which supports decisions on limits of water withdrawal from the Amudarya river by states and water consumers, for consideration at ICWC meetings. This is accomplished in the following way:

- The region's states submit a proposal on ten-day water limits by oblasts for one or another period of time to BWO Amudarya 10–15 days before the respective period starts.
- On the basis of submitted requests, BWO Amudarya comes to an agreement with oblast water management organizations about water withdrawal limits for every water consumer.
- Summarized information on water withdrawals by states (national limits) is submitted for consideration and approval at regular ICWC meetings.
- Upon approval of the national limits, BWO Amudarya updates, adjust and approves ten-day limits on water consumers for one or another period and passes the limits to its territorial branches for execution.

It should be noted that the following options for water allocation are selected at ICWC meetings, depending on forecasts and the actual water situation:

1. In the period of normal water supply and availability of water resources in reservoirs, water distribution is to be carried out according to the set water withdrawal limits.
2. In low water periods, Article 4 of the Almaty Agreement of 18 February, 1992 is applied, which stipulates the following criteria for the interstate use of set water withdrawal limits:

- at water availability lower than design one, water withdrawals on the ICWC decision should be reduced proportionally as applied to the whole river basin;

- maximum amount of permitted over-withdrawal of water is set at a level of not more than 10%, as applied to single periods;
- the criteria for introducing water shares (in %) is a water resources deficit.

At the same time, the share of water withdrawal is determined on the basis of approved ICWC limits for the whole period, as applied to major water users.

Under conditions of normal water availability in the basin, there are no particular problems in managing and allocating surface waters. In low-water periods, however, water management is complicated. This is especially so in extreme cases, where, despite double control by BWO and ICWC, decisions made by ICWC on the reduction of water withdrawal limits, and held as obligatory for all water consumers, are not always observed. This is caused by a number of factors, such as an imperfect legal and regulatory base for ICWC's executive organizations, differences in national interests regarding water use, and poor equipping of BWO organizations with water meters and monitoring facilities. As a result, decisions made by ICWC have not always been implemented in accordance with set volumes and time limits.

Some problems emerged under extreme low-water conditions in 2000–2001 and at the beginning of 2002 that are deserving of special attention:

1. Inequity of water resources use along the river arising when upstream water users divert more water than downstream users.
2. Increase of non-productive losses on all river reaches.
3. Breach of water discipline by water users, especially relating to intake by pumping stations.
4. Interference of local authorities in day-to-date water control.

Water is a key factor of social and economic well being in Central Asian countries. At present, the region is generally not experiencing acute water shortages, although the demands of the populations are met largely at the expense of environmental needs, which are practically neglected. During certain periods of time in some zones of the region (especially downstream the major rivers), water shortages (both in terms of quantity and quality) have already impacted upon the socio-economic situation. In the near future, water shortages are likely to increase, particularly bearing in mind factors such as climate change. Such increased shortages will jeopardize sustainable development in the whole region.

The main water problems in the basin are concentrated in the downstream area (this is a primary feature as concerns water resources management in the Amudarya river basin). This area suffers from acute water shortages during normal or dry years, and lacks the flow necessary to conserve natural ecosystems and to restore part of the Aral Sea. Despite the efforts made towards the equitable allocation of water resources between water consumers, even inside one country there has been difficulty in avoiding disproportionate and inequitable water consumption, especially between midstream and downstream areas. To solve this problem, we need to develop effective management mechanisms and rules that would counter flow losses and promote sustainable water allocation, including the provision of water for in-stream needs specifically during low-water years.

Specific steps have been taken by BWO Amudarya to mitigate shortages of water:

1. Upon agreement with the basin's states, the BWO has proportionally reduced water withdrawal limits.
2. During the whole low-water period, the BWO actively cooperated with the Ministry of Water Resources of Turkmenistan and the Ministry of Water Resources and Agriculture of the Republic of Uzbekistan regarding the possible reduction of water withdrawal in the middle reaches, so as to allow water to pass to the downstream area. This activity produced some positive results.
3. During the whole low-water period, because of low inflow to Tuyamuyun waterworks and a lack of regulation of water volume in reservoirs, in the downstream area water was allocated proportionately among three main downstream water users (Dashoguz oblast, the Republic of Karakalpakstanm, and Khorezm oblast), on the basis of inflow to Tuyamuyun waterworks. The water allocation was based on decisions made during routine meetings of the commission concerning the development of the required mode of operation of Tuyamuyun waterworks in case of water shortages.
4. In order to increase control over observance of the set water limits the BWO, together with the Ministry of Water Resources of Turkmenistan and the Ministry of Water Resources and Agriculture of the Republic of Uzbekistan, created the control operational groups on river reaches that operated during the whole low-water period. Violators of water discipline were subject to punishment such as notice, penalties, and dismissal. Such strong control over water withdrawals also had some effect as concerns the maintenance of water discipline and provision of inflow to both Tuyamuyun waterworks and Takhiatash waterworks, being the end station on the Amudarya river.
5. The BWO, in cooperation with Uzbekglavgidromet (Uzbek Hydrometeorological Service), made measurements at Tuyamuyun and Kipchak hydroposts, located in the lower reaches. In the middle reaches, the BWO made measurements at key hydroposts together with the representatives of Turkmengidromet (Turkmenistan Hydrometeorological Service).

6. The BWO commissioned work on achieving a more precise definition of water losses in the reach from Tuyamuyun waterworks to Takhiatash waterworks.

Despite these measures, however, we failed to avoid the negative consequences of water shortage in the region as a whole. As a result, downstream water users have suffered most of all. Though the above mentioned emergency measures on provision of flow to downstream areas and on uniform allocation of water along the whole river channel had some positive effects, in general the water-related situation has not been changed. Indeed, although steps taken by the states to create an integrated water resources management system and to strengthen regional cooperation in the Amudarya river basin have produced some positive results at certain stages, the problem of integrating water management in the basin has not been completely solved.

Under the present political and socio-economic conditions, one of the most reasonable ways to create prosperity in the region is to implement the principles of integrated water management aimed at effective water use and reduction of unproductive water losses by developing all-round cooperation. To organize such regional partnerships, we need to integrate our efforts in the following directions:

- Promotion of integrated water management and conservation by developing partnership at interstate level.
- Integration of the interests of the economic sector and the environment through inter-sector partnerships in each state, where environmental needs would take on key importance
- Integration of knowledge and practices through the partnership of science and industry with water users and water management organizations (by using such tools as knowledge base, training, extension and innovation services)
- Closer co-operation between international donors and the region through coordination and partnership of international funding agencies.

The urgency of ensuring an effective mechanism for joint water management in the Aral Sea basin was specifically noted in the Decision of the Heads of Central Asian states on "Major Directions of the Program of Concrete Actions on the Improvement of Ecological and Socio-Economic Conditions in the Aral Sea Basin for 2003–2010," where BWO's interests were considered. The result was the creation of the institutional framework of IFAS in the Aral Sea basin. If such a framework is exploited to its full potential, it can ensure sustainable interstate water management in the region, as well as effective and rational water use. In order to make the IFAS framework effective, the states in the region need to carry out the following priority measures aimed at the intensification of interstate cooperation in water management:

1. Sign the Interstate Agreements on the separation of functions in water management and use and on joint trans-boundary water planning. Those agreements would allow the regulation of water management system functions.
2. Sign a number of Interstate Agreements dealing with such issues as the institutional framework of joint water management, information exchange in the region, and limits of water inflow to the Aral Sea and Priaralie.
3. Augment the status of ICWC.
4. Strengthen the role and status of BWO "Amudarya."
5. Sign a single Agreement on the basic principles of joint use of trans-boundary waters in the Amudarya river basin.

IWRM FINANCIAL, ECONOMIC, AND LEGAL ASPECTS: THE EXAMPLE OF THE "IWRM-FERGHANA" PROJECT

A. Jaloobayev
Water Department, Kyrgyz Republic

As a solution to the global water crisis, *Integrated Water Resources Management (IWRM)*, combines management based on hydrographic principles, comprehensiveness, recognition of different interests, coordination of all hierarchic levels, consensus, interdependence, and wide involvement of water users. Integration suggests interconnection between natural systems—water resources quantity and quality—and the social systems that define water demand and disposal, pollution control and social sustainability.

According to the IWMI definition, *IWRM is* [a form of] *governance, financing water services under coordinated participation of all water users within the clear framework of water policy, laws, institutions and technologies and as result optimization of social and economic water value, its use efficiency and distribution uniformity under simultaneous important ecosystems' sustainability.* The Global Water Partnership, meanwhile, defines IWRM as "[a] *process facilitating coordinated water, land and other related resources management and development to reach* [the] *maximum economic and social level on* [an] *equitable basis without damage to vital ecosystems sustainability".*

The concept of IWRM is emerging as a solution to growing water issues, and is topical both for Central Asia in general and the Ferghana Valley. Major issues are as follows:

- Growing food demand as a result of population growth;
- Growing competition between agricultural and other subsectors (e.g., hydropower engineering);
- Inadequate access to safe drinking water and sanitation;
- Inadequate or worn infrastructure;
- Low income and declining employment opportunities;
- Low land and water productivity;
- Deterioration of quality of water and environment (e.g., from soil degradation, water logging and land salinization);
- Parochial interests;
- Management disconnection and lack of appropriate policy.

IWRM is a process which provides coordinated strategic development and efficient management and use of land, water and other resources, in

P. Wouters et al. (eds.), Implementing Integrated Water Resources Management in Central Asia, 45–53.

combination with technical and scientific innovations, to maximize economic and social well being based on principles of equitability and limitation of damage to vital ecosystems. Such programs aim to overcome:

- Lack of coordination as a result of national and administrative restrictions;
- Sectoral restraints;
- Breaches in hierarchic levels of water management;
- Lack of interest of water users and public on the one hand and managerial bodies on the other hand, resulting in impossibility of open discussion of all issues;
- Lack of interest of all participants (and leaders) of water management hierarchy in final productivity of water resources;
- Weakness of personnel provision and its training;
- Gap in technical level of equipment and structures;
- Instability in financing of operation and development.

The participation of all concerned parties in water resources planning, development and management process is indispensable condition for successful IWRM implementation.

In Ferghana Valley, there is significant disparity between water demand and supply. Most of the water in the SyrDarya basin comes from Kyrgyzstan and Tajikistan, although most irrigated land is located in Uzbekistan. Whilst 75% of the flow formation of the SyrDarya is in Kyrgyzstan, the country has only 20.6% irrigated lands and receives 18.4% of total water supply to the valley. Uzbekistan has 62% irrigated lands and receives 69.7% of total water supply. Tajikistan has only 17.3% irrigated lands and uses 11.8% of water supply. This creates tension in water distribution between the countries. There are substantial discrepancies between water demand and supply, actual water intake, and established limits.

Most small and medium rivers and canals in the Ferghana valley cross country, oblast and rayon boundaries. Water organizations have mandates only for water for irrigation. Industrial, municipal, and other water use is the responsibility of other sectors. Administrative-territorial and sectoral non-coordination prevents effective, uniform, and productive water management.

Efficiency of water transportation and use in the Ferghana valley irrigation systems amounts to 45%, two thirds of which takes place in tertiary canals. The main reason for low systems efficiency is their bad state and poor irrigation technique. Significant changes in water demand and supply relations are observed along, and between, canals.

The Ferghana Valley encompasses three oblasts in the Kyrgyz Republic (Osh, Jalalabad, and Batkent); one oblast in Tajikistan (Sogd) and three more in Uzbekistan (Ferghana, Andizhan, and Namangan). The main goal of the project is the development of recommendations directed at realization of

IWRM principles in selected pilot canals in the Ferghana Valley. Such recommendations would support social harmony during the complex period of agricultural reform.

The following canals were selected as pilot sites: South-Ferghana (Uzbekistan), Aravan-Akbura (Kyrgyzstan) and Gulyakandoz (Tajikistan). The main areas selected for the inception phase were South-Ferghana Canal in Ferghana, the Andizhan oblasts in Uzbekistan and Osh oblast in Kyrgyzstan, the Aravan-Akbura Canal in the Osh oblast of Kyrgyzstan and the Gulyakandoz Canal in the Sogd oblast of Tajikistan.

Shortcomings in existing water resources management system	Possible results under transition to IWRM
Aspects of water policy	
Lack of clear water policy meeting modern requirements in water resources management and use.	Effective integrated policy under IWRM, providing effective financing of water sector and directed at water conservation and land and water productivity increase.
Discrepancy between water policy, water legislation, management and procedures of water resources management and use.	Integration of water policy and legislation, organizational needs, water resources management and use.
Preservation of state order and fixed purchase prices on major agricultural products (fully in Uzbekistan and partially in Tajikistan).	State makes water use free. Crop pattern is determined by market demand and technology, and prices are regulated by the market.
Legal aspects	
Discrepancy between legal status of water structures and their functioning under transition to IWRM. Strict top-down procedures of water management.	Legal changes connected with status of water structures facilitating management changes and providing independent decision making under society democratization and transition to market economy.
Sectoral aspects	
Disconnected executive bodies lacking coordination (rural water supply, industrial water supply, power engineering and other consumers).	Establishing of inter-sector coordination providing integration of actions in water resources management and use (through canal, system, subbasin, basin Water Councils).
Administrative borders of management (parochial egotism in water resources use).	Management on base of hydrographic borders, warranty of stable and equal water supply independently from user location.
Lack of incentives for water organizations and users to increase water productivity.	Recognition of water as limited resource. Introduction of incentives for water productivity increase and water conservation in practice of water organizations and water users.

Non-***participation of many concerned parties in decision-making process. Lack of accountability before water consumers.	Public consultations, involvement of concerned parties in decision-making process. Management participation of all concerned parties; services supplied on contractual basis.
Technical and technological aspects	
Significant water infrastructure depreciation and need of rehabilitation.	Creation of conditions for water objects rehabilitation through water services market, as result of water conservation under economy stabilization.
Disadvantages of water metering services and need in equipment. In this connection, uncertainty in actual water expenses.	Conditions for water metering services perfection at expense of certain financial sources. Clear account of water diversion and disposal.
Lack of automation, efficient control and communication means.	Finding investment to solve problem.
Unevenness in water resources distribution, lack of water supply stability and equitability.	Conditions for even and equitable water resources distribution and water supply to users.
Significant organizational losses due to non-coordination of different hierarchic levels.	Minimization of organizational losses through clear coordination of all management levels.
Lack of common account of surface and ground water use as well as return flow utilization.	Common planning surface and ground water use with return flow utilization.
Economic aspects	
Lack of funding from budget for water infrastructure O&M and rehabilitation.	Possibility of water organization funding stabilization at expense of self-financing under partial state support to development.
Lack (Uzbekistan) or imperfect (Kyrgyzstan, Tajikistan) paid water use. Uncertainty of actual financial cost of water services. Lack of connection between services and payment.	Conditions for paid water use in Uzbekistan and its perfection in Kyrgyzstan and Tajikistan. Planning tools and charges based on actual cost of management. Realization of principle of "payment for service". Mechanism of services recovery.
Lack of progressive forms of water users organization (WUA, WUAF) and non-coordination (legal and economic) of water users.	Water users associations have clear rights and obligations in relations with water bodies. WUA and WUAF integration to joint water management under IWRM principles.
Ecological aspects	
Lack of attention to water pollution and harmful impact issues and environmental issues in general.	Rising general level of environmental protection and water limit allocation for this purpose.

Based on world experience in IWRM implementation, and analysis of the situation in the Ferghana Valley within the project area, the following IWRM vision for the Ferghana Valley was formulated:

1. Water management with regard to the range of political, social-economic and administrative systems under IWRM will allow the provision of stable water services on basis of participation and within the framework of policy, legislation, organizations, and technologies.
2. Water organizations' jurisdiction will be based on hydrographic borders. Such a measure will permit administrators to make timely decisions and provide water services without political intervention.
3. Water will be managed in an integrated manner with regard to all types of water use within a single hydrographic area. Hydro-Meteorological information about dynamic water supply and multifold utilization will be available for all water users in a practical format.
4. Strategic planning of water use will be based on agricultural production planning, and municipal, rural and industrial water supply. Water's social and economic value will be optimized and its efficient use and equitable distribution will be provided for.
5. Decision making in water management will be decentralized at a lower level (WUA and its federations, Canal Council) and based on a regulating structure defined by national legislation. The Government will support the establishment and functioning of WUAs and their federations.
6. Government will shift from direct management to water basin and subbasin management and water sector regulation; water organizations will have a clear mandate in certain management functions within their jurisdiction. Their level of competence will be defined by the regulating structure.
7. An elected council, including water users and stakeholder representatives, will govern each WUA. Governing boards will have a clear mandate regarding policy, rules and procedures in their water systems, within the national regulating structure.
8. Farm income will be so satisfactory that farmers will be able to pay for a system of canals O&M. Irrigation and water supply systems funding will be based on actual expenses and state subsidies assessment. Effective arrangements, incentives, monitoring and auditing will ensure the accountability of water organizations. A dispute resolution body with a clear mandate will be established at WUA level in the Ferghana Valley and SyrDarya Basin.
9. Water organization's (including WUA) governing boards will take part in BWO rules formulation on transboundary issues. Clear legislation will be formed gradually for transboundary agreements within national and transboundary basins.

The main goal of effective water management in the Ferghana Valley is reliable water supply for all water users, with irrigated farming being the priority. The main indicator of effective management is even water distribution among water users and timely water supply to each water consumer. The criterion of effective water management is irrigation water productivity.

To achieve high water productivity it is necessary to create a system of water accountability at all hierarchic levels and to define crop water requirements. It is necessary to take into consideration water scarcity in the region, because of which equitable, sustainable and rational water allocation is very important. All water management levels are interconnected and interdependent.

Wide involvement of HGOs and public opinion in the management process is also very important, including prospective planning and projects, water allocation technology and conflict and disputes resolution between different structures and water users. Water users should actively participate in public opinion formation.

1. IWRM Legal Aspects

Legislation lays a foundation for competent use, definition of responsibilities and rights to establish necessary institutions and mechanisms for policy implementation. Legal provisions stipulated in 'Law on water', 'Water Code', or 'Law on land' and other provisions should accompany effective state policy in water resources through the definition of the role and responsibility of the state, water bodies, and other parties involved in water distribution, use, management, development, and protection. Important factors include:

- Clear definition of water's social, economic, and ecological value;
- Creation of concrete positions on restructuring, competence, privatization, strengthening local communities, and water user's participation;
- Clear definition of water rights, WUA role, rules of coordination between sectors and its mechanism;
- Establishment of links with environment, agriculture, local authorities, economic development, etc.

Water legislation defines water ownership and water rights status and terms, as well as the base for its distribution. Legislation also creates mechanisms and incentives for irrational water use limitation, reduced pollution, and cost and benefits distribution within society. Legislation establishes mechanisms for dispute resolution and charges society with obligations on water resources protection, IWRM funding, etc. Legislation may also foresee the establishing of executive agencies for the coordination of activities in various sectors. It often includes the participation of national bodies at the highest level in policy formulation, planning, and coordination (for example, the national water council). Legislation also regulates monitoring, management, and information exchange.

The legal base is an important part of establishing a WUA, allowing the creation, improvement, and stimulation of WUA and water users activity in the following issues:

- WUA water rights;
- Establishing water market from saved water within WUA;
- Soft credit and privilege taxation;
- Sanctions for water use rules breaches both for water users and water organizations;
- Recommendation of certainty in ownership of on-farm objects under WU establishment;
- Legislating for tax allowances, e.g., within Ferghana Valley, VAT exemption for WUA;
- Simplifying scheme of WUA registration;
- Ordering of sanctions for water use rules breaches by water organizations.

Within the project framework, existing water legislation analysis has been performed including water use in irrigated agriculture, water funds, and taxation in countries within the Ferghana Valley. To perfect the WUA legal base, there is the possibility of using the Kyrgyz Republic's "Law on WUA" as a model juridical tool. Special attention was paid to WUA water rights, water market creation, ownership of irrigation systems and WUA property, and privilege taxation. As a result, recommendations and proposals on adjustments to existing legislation were developed. WUA draft statutory documents were prepared with regard to legislation in Tajikistan and Uzbekistan and draft agreements were draw-up regulating current WUA activity between the WUA and farmers regarding paid water services.

Typical conflicts and disputes, such as those between water users in farms that have primary and secondary water users, between water users within the WUA, between the WUA and water users, and between the WUA and water organizations, were studied. Existing mechanisms for dispute and conflict resolution were considered with regard to current civil legislation and appropriate procedures. In other words, analysis was made of the three countries' legal base in connection with the system of justice, civil legislation, arbitration and civil processes. On the basis of this analysis, recommendations were given on disputes and conflict resolution.

2. ECONOMIC AND FINANCIAL ASPECTS

Water sector financing has been influenced by many factors, including the collapse of the USSR, reductions in cotton price, yield and agricultural income, electric energy price increases, and the weak market and state in Uzbekistan. Uzbekistan can provide less than 50% of the necessary funding for the main and secondary canals O&M. In Kyrgyzstan and Tajikistan, the situation is even worse because of the charged water delivery system, whereby water charges are shared between the state and water users. Lack of cash and the widespread practice of barter shows that Uzbek farmers are unlikely to cover the full cost of the irrigation network O&M without additional economic liberalization and growth.

Presently, according to existing legislation, different funding systems take place in the three states located in the Ferghana Valley:

- In Uzbekistan, because of the absence of paid water use, water structures at all levels are funded from the state budget. Water charges are likely to be introduced in the future.
- In Kyrgyzstan and Tajikistan, where there is paid water use, funding is shared between the state and water users.

In the future, conditions of economic liberalization and open access to markets will mean that farms will be able to cover O&M costs up to the level of distributive canals and, in time, beyond that to the whole irrigation system. Financing of water services of the upper hierarchic level will be based on actual costs, shared between water users (WUA, WUAF) and governmental bodies, and allied to effective measures and incentives for efficiency.

The financing of water structures at basin level in the Ferghana Valley, under the active participation of all water users, will consist of following functions:

1. Assessment of expenses for water organization regulation and governing activity at subbasin level;
2. Assessment of expenses for water services;
3. Draft budget development;
4. Definition of sources of financing, including necessary fees, local charges, and payment or taxation for services;
5. Definition of base for fees, local charges and their collection, allotment and use.
6. Necessary procedures of financial accounting and auditing.

The financing of water services at the level of separate canal systems, under the active participation of all water users, will consist of the following functions:

1. Assessment of expenses for necessary WUA and WUF governing activity at the level of main, secondary, and tertiary canals (including meetings, travel cost, auditing, etc.);
2. Assessment of expenses for necessary water services at each hydrographic level (including irrigation, drainage, fees collection, administration, etc.);

3. Development of WUA draft budget at each hydrographic level;
4. Definition of source of funding including water service charges and possible state assistance;
5. Definition of base for charges (per hectare, water supply unit, agricultural crop) as well as their collection, allocation, and use;
6. Necessary procedures of financial accounting and auditing.

One further financial problem is WUA taxation. Recommendations on making the WUA free of taxation were made within consideration of WUA activity legislation in the Ferghana Valley.

The WUA economic situation depends largely on main assets and irrigation system transfer. The WUA should replace the former structure of on-farm irrigation network O&M because agricultural reform means that a lot of small farms have appeared in place of larger ones.

An important question is the condition of secondary canals and other objects funded by MAWR transfer to WUA. This can be solved in one of two ways:

1. During the first 5 years after the WUA has been established, water objects would be transferred to the WUA on a contractual basis for provisional use, along with annual money transfers to cover the cost of O&M. Water objects transfer for provisional and permanent operation is foreseen both in Uzbekistan (Chapter 31, "Law about water and water use") and in Tajikistan (Chapter 10, Water Code).
2. The State water-related organization could become the founder of the WUA. Water objects transferred to WUA become the state's contribution. Moreover, the water organisation, as a founder, would be responsible for covering the cost of these objects O&M within normative needs.

Further realization of the IWRM-Ferghana project will show the necessity of further amendments and additions to national legislations.

ENSURING OF STABILITY AND EVEN WATER DISTRIBUTION AT NATIONAL AND LOCAL LEVELS

A. I. Tuchin

1. Introduction

According to the results of the analysis carried out at the initial project phase, for IWRM–Ferghana system, a hierarchical water and land resources management structure reflecting both the present water relations and information flows formed within the framework of the project being implemented was recommended. Ferghana Valley basin management level including *BWO* information contacts was accepted as an upper management hierarchy level. At the second hierarchy level, there are large irrigation systems and separate canals (*LSC*) that are subordinate to *BWO* management level and provide bulk transfer and distribution of water resources among diverse *WUAs* and *private farms*. The following hierarchy level is Water User Association (*WUA*), defined as an independent water management unit that sets water resources limits on water intakes and controls water supply in the context of canals and administrative areas. As a lower management hierarchy level, separate *private farm* level was accepted where crop pattern and soil salinity conditions, irrigation technique and technical condition of on-farm irrigation and collector-drainage network were identified.

The concept of multi-level water distribution control structure provides consecutive achievement of strategic and tactical goals, and distribution of water resources among participants in yearly, month, decade and daily context, at that each lower level follows control strategy set at higher hierarchy level, except its own effectiveness criteria.

2. Irrigation Systems Characteristics

The objective of any irrigation system is to transport required bulks of water to given sites and in given time. That is why the technique for water resources distribution, at each hierarchy level, relies on own set of hydraulic structures, volumes of water resources supplied from higher level, and requirements from lower objects. The main operational performance characteristics of irrigation system are:

- availability of water for irrigated areas and water users;

P. Wouters et al. (eds.), Implementing Integrated Water Resources Management in Central Asia, 55–80.

- evenness of water distribution among all participants;
- evenness in time of water supply to participants;
- volumes of water resources losses in system;
- system controllability.

Considering irrigation system as an object of control, each of the above-mentioned characteristics can be transformed into individual (private) criterion, reflecting system performance, i.e., of selected criterion. Using various criteria will result in various options of water resources distribution. Therefore, the task of irrigation system control is among multi-criterion tasks, at which the uncertainty of objective requires introducing additional hypotheses on significance of this or another characteristic and individual criteria allowing for linear (or nonlinear) convolution. The uncertainty of objectives is not a single form of uncertainty in irrigation system control. The next form uncertainty is related to stochastic nature of hydrological flow, of which available volumes of water resources form, as well as to weather conditions, under which specific requirements of water users form. This form of uncertainty, usually called "natural uncertainty," precludes from finding a single solution, and requires seeking some strategy for water resources distribution control as a function from inflow volume and current weather conditions.

For integrated assessment of irrigation system performance by each individual criterion, it is necessary to formulate a set of indicators adequately reflecting dynamics of this criterion in both the system as a whole and its separate elements. Thus, any indicator is not an independent element, which can be used without relatively selected criterion. Undoubtedly, in intricate technical systems where spaces of objectives cross, situations when one indicator can reflect dynamics of several criteria at once arise, but this is only an indirect reflection, which can be used in direct calculations on no account, since differences in normalization conditions of different criteria can lead to unexpected results. In engineering practice, the so-called set of "check indicators" [1] is often used as a means to overcome the uncertainty of objectives. This tool (the simplest but not the best) is applied when numeric restrictions are put on some of indicators. Restrictions can be two-sided or one-sided, it is not so important; the main simplification is that now only one criterion remains, by which system performance is assessed, i.e., the solution becomes univocal. The main drawback of this method is that for the vast majority of real objects, it leads to mathematically unsolvable problems. Physically, it will only mean that some of system parameters would exceed the acceptable bounds, and mathematical condition of the system would be indefinite.

Sets of indicators being used, at each hierarchy level, can be different; furthermore, they can change at diverse stages of system control, depending on the present importance of this or another criterion. Only those compositions of indicators, through which inter-hierarchy coordination of water resources distribution is provided, require compulsory matching in normalization. Before going on to selecting and basing indicators for different hierarchy levels and different stages of control, it is necessary to state some general principles, to which one should orientate in developing them for irrigation systems:

(a) Any set of indicators, introduced to assess any of irrigation system performance characteristics, should have no less than one algorithm that ensures calculating an integral value of indicator by the considered characteristic, for the entire system;
(b) Indicators should be uniquely calculated through variables, characterizing the status of irrigation system objects;
(c) Indicators should be scaled by any most important parameter of irrigation system in order that there be an opportunity to compare operational characteristics of different irrigation systems,
(d) Values of indicators for separate irrigation system elements should reflect a real proportion of their loading, in system performance.

The last requirement is related to the fact that values $\sim 500 \div 1{,}000$ periodically arose in calculations of indicators on pilot canals, for individual elements. As the further analysis showed, these values did not correspond to that the given hydraulic structure was loaded $500 \div 1{,}000$ times more (or less) than the norm, but only pointed to failure of the selected scale of used indicators. Regardless of the triviality of points (a), (b) and (d), the most part of indicators proposed in the project does not meet the above-mentioned requirements.

Besides the separated requirements, the issues related to size of time intervals, during which numeric values of indicators keep their worth, stand quite particularly; as everybody knows, the value of any information falls as time goes by, at that the rate of fall in this "value" strongly depends on type of selected indicator.

In mathematical analyses of control processes at the existing irrigation systems, in addition to the problem of adequate formal description of physical elements of the system and their influence on each other, there is a problem related to proper description of the current relations between its diverse participants (at newly formed systems there is usually no problem like this). The importance of the second problem becomes more apparent in calibrating phenomenological coefficients of mathematical models, because it requires exact formalization of present rules for water resources distribution among participants. But just these rules are the most veiled elements of irrigation system that generate one more type of uncertainty. Under these conditions, researcher, developer of mathematical model, faces inevitable problem of choice between coarsening a model and, on the contrary, detailing it. In the first case, certain distortion of physical and technical indicators of the system is knowingly allowed so as to identify existing relations between participants while identifying parameters, in an implicit state. Such models gave a good account of themselves in production control if external environmental conditions do not undergo sudden changes. The second way is based on more detailed description of physical processes and technology of system elements so as to identify parameters by indirect estimations or even by other similar systems. In this case unknown relations do not participate in identification of parameters, and are determined at the second stage—identification of relations or rules for water resources distribution that exist in the system, at considered time. Though the second way is very laborious, the results achieved on such models not only enable to manage the system in wider range of change in external conditions, but also give opportunity to identify separate social regularities having independent value. In both the cases, it is necessary to develop a set of indicators providing calibration of models for identification of model parameters These indicators do not relate to system performance efficiency indicators [2], and reflect cognitive side of researches more than technological one.

3. Main Stages in Water Distribution Control

In control of water resources distribution process as well as in any technological process, separate periods, stages, tasks differing in time intervals and character of information used for finding solution can be marked out. In this work, the issues of reconstruction and forward planning of irrigation systems are not considered, so the largest period of control is time interval equal to one year (hydrological year), which consists of two periods—vegetation "T^V" and non-vegetation "T^N." In its turn, each period is divided into time intervals equal to one decade and marked with "t." To mark time within a decade, sign "τ," $\tau \in t$ is used. We will mark considered time interval with "t," and previous and next ones respectively with "$t - 1$," "$t + 1$." Technical characteristics of system elements keep their values during the entire control period. The latter admission shows that emergency situations in irrigation systems are not considered at the given stage of analysis, indirectly they can be considered as reduction in parameters for characteristics "volumes of water losses in the system" and "system controllability."

According to the existing terminology, yearly period of water distribution control consists of the following stages (Figure 1):

- annual planning
- operational planning
- operational control.

Annual planning is carried out once for vegetation and non-vegetation periods.

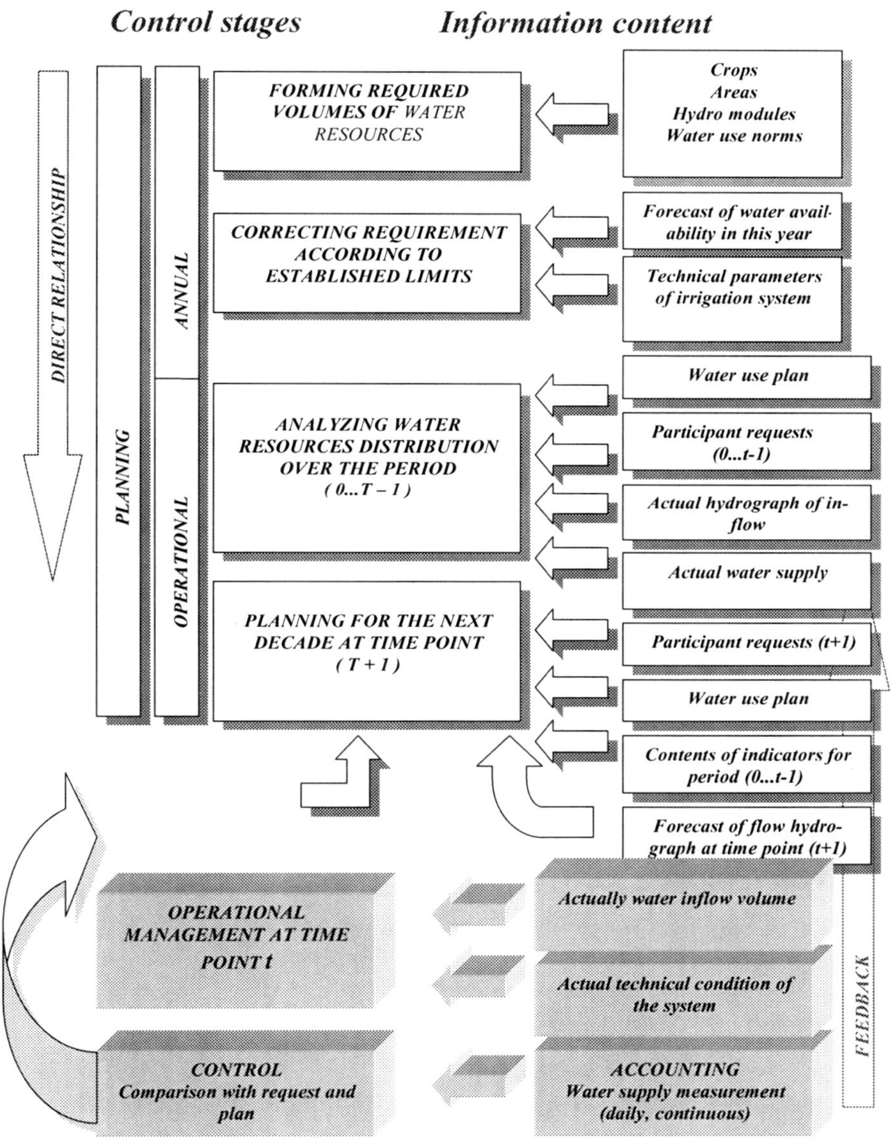

Figure 1.

Task 1—Determining volumes of water resources required by participants, *Initial information content:*

– contents of crops,
– areas under crops,
– allocation of areas in hydro module districts,

- norms of water use by crops,
- leaching norms for irrigated areas,
- mean annual hydrological characteristics of district.

Solution methods:

- count-up, optimization arrangements are not practically used.

Solution results:

- required volumes of water for each participant for the whole period and in a decade context.

Indicators of solution quality assessment:

- specific water productivity is average weighted for the territory ($/m$^3 \times$ ha)
- water productivity in the context of crops (m^3/ha).

Task 2—Correcting volumes of water resources according to the established limits.

Contents of initial information:

- forecast of water availability in the current year,
- technical opportunities of irrigation system.

Solution methods:

- resources allocation in the network with a limited capacity, optimization type.

Solution results:

- allocated water volumes for each participant for the entire period in a decade context.

Indicators of solution quality assessment:

- specific water losses in system (efficiency),
- evenness of curtailing participants in annual volume of water resources,
- evenness of curtailing participants in decade volume of water resources.

The result of solution for both tasks of annual planning:

- corrected water use plan.

Annual planning is carried out in two stages, first in non-vegetation period, or water resources allocation in vegetation period, then in vegetation period for water resources allocation for non-vegetation period in the next hydrological year, the sequence of tasks for these stages is given in Figure 2.

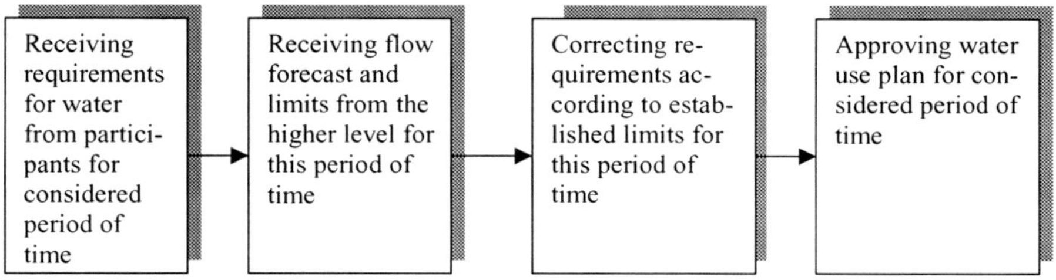

Figure 2. Annual planning.

Operational planning is carried out every decade, "t"—number of decade.

Task 1—Analysis of water resources distribution for the past period $\{0 .. t-1\}$.

Contents of initial information:

- water use plan,
- requests of participants, for a period of time $\{0 .. t-1\}$,
- actual water supply to participants for a period of time $\{0 .. t-1\}$,
- actual hydrograph of inflow for a period of time $\{0 .. t-1\}$.

Solution methods:

- statistical methods.

Solution results:

- contents of indicators by chosen criteria for a period of time $\{0 .. t-1\}$.

Indicators of solution quality assessment:

- statistical assessments of indicators straggling.

Task 2—Correcting volumes of water resources supplied to participants at time point "$t+1$."

Contents of initial information:

- contents of indicators by chosen criteria for a period of time $\{0 .. t-1\}$,
- corrected water allocation plan for time interval "t,"
- requests of participants for time point "$t+1$", which are made according to actual dates of crop sowing and irrigation as well as to current weather conditions,
- forecast of flow hydrograph for time point "$t+1$."

Solution methods:

- gaming methods, optimization methods (determinate, stochastic).

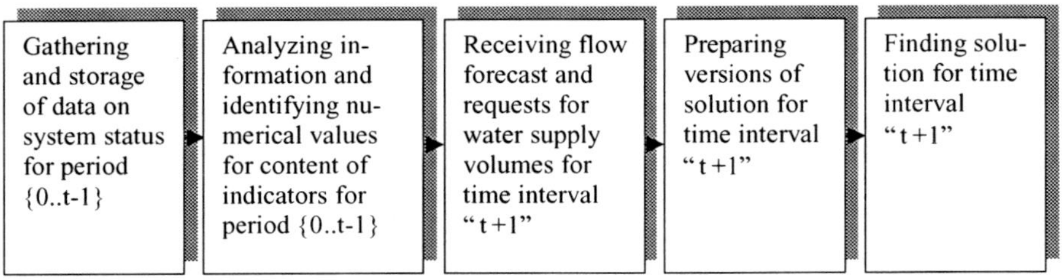

Figure 3. Operational planning.

Solution results:

– volumes of water resources supplied to participants at time point "$t + 1$."

Indicators of solution quality assessment:

– reduction in statistical straggling of indicators.

The sequence of task solution at the stage of *operational planning*, for time interval "t," from receiving information to implementing solution is given in Figure 3.

Operational control is carried out every day (hourly).

Task 1—Supplying volumes of water resources to participants according to values established at the stage of operational planning, for time point "t"

Состав исходной информаиии

– decade volumes of water resources for each participant,
– actual hydrograph of inflow at time point $\tau \in t$,
– actual status of hydraulic structures.

Solution methods:

– regulation methods

Solution results:

– daily (hourly) volumes of water resources supplied to participants.

Indicators of solution quality assessment:

– statistical straggling of hourly and daily flows in relation to the given.

Task 2—Accounting supplied volumes of water resources.

Content of initial information:

– actual measurements of inflow at time point "$\tau \in t$,"
– actual status of water-measuring facilities at time point "$\tau \in t$."

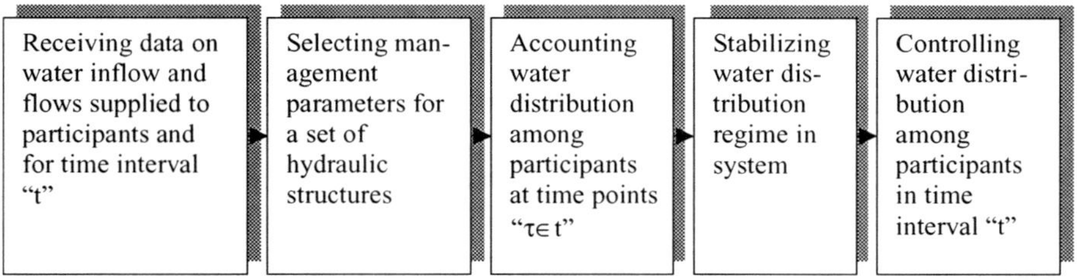

Figure 4. Operational control.

Solution methods:

– interpolation methods

Solution results:

– actual measurements of water flows for each participant
– actual measurements of water levels along canals.

Indicators of solution quality assessment:

– range of fluctuations in separate measurements and errors of water-measuring facilities.

At that τ – time point within decade.

Operational Control stage provides implementing solutions found at the stages of Annual and Operational Planning. The sequence of solving tasks at *Operational Control* stage is given in Figure 4.

Solving *Task 2, Operational Control*, forms a feedback loop necessary to solve tasks at *Operational Planning* stage.

4. Analysis of Water Distribution Control Process

4.1. ANNUAL PLANNING

Individual tasks of each stage in irrigation system control have different information base for diverse objects and hierarchy levels. For example, information base of the task "correcting volumes of water resources according to the established limits" at *Annual Planning* stage is principally different for those *LSC*, at the head of which regulating reservoirs ensuring compliance between required and supplied hydrographs are located (in this project such objects are SFC and Aravan–Akbura Canal), and *LSC*, the head of which is fed by natural inflow (Gulyakandoz Canal). In the first case, flow hydrograph is formed at basin level depending on possibilities of regulation and guarantees of observing limits, which "..., in the example of Syrdarya, often depend on policy and discipline of basin management rather than hydrology and regulation degree[1] ...." In the second case with non-regulated flow, it is also necessary to determine value of guaranteed volumes, but now on the assumption of probabilistic curve of concrete stream availability. Thus, in both cases, water use plan should be based on some *guaranteed flow values*, and actual forecasts of water availability should be regarded as possible additions. Of course, for objects with non-regulated flow, the role of *Operational Planning* stage rapidly raises. Furthermore, the significance of information on average decade volumes descends as hierarchy levels lower, because the nearer we approach to field, the narrower time interval, during which water is supplied, becomes; for separate water outlets, it (water supply time interval) is calculated in days. Under these conditions, tasks of *Operational Planning* and *Operational Control* stages cross, since the idea of average decade flows losses meaning.

This minor analysis points to the necessity of dividing the content of indicators by hierarchy levels and head water intake feed conditions. Moreover, at present the most important step is to form indicators, which do not lose meaning in transiting from level to level and from system to system. In addition, no less important characteristic of indicator is time interval length, during which its numerical value keeps its worth, so a set of indicators based on any integral characteristic of irrigation system will have an advantage. The following characteristics are more evident for *Annual Planning* stage:

- *availability of water for* participants *over control periods*
- *specific water supply* (m^3/ha)
- *specific water productivity* (\$/m^3)
- *volume of water losses in system.*

Thus, a set of indicators should consist of:

– *relative availability of water for* participants *over control periods*
– *relative specific water supply for participants*
– *relative water productivity in system*
– *relative volume of water losses in system.*

As basic values of these indicators, their *guaranteed values* calculated by multi-year series are used. For years strongly differing from average annual in water availability, additional characteristic "system controllability" can be

[1] V.A. Dukhovny, comments to this work.

recommended to include in the set of indicators. For this stage, the following questions remain unsolved:

- *Do curtailing rules depend on year water availability?*
- *Do all participants have equal priorities?.*

4.2. OPERATIONAL PLANNING

It is seen from the information content of *Operational Planning* stage that information flows for diverse time intervals are used for tasks of correcting volumes of water resources supplied to participants at time point "$t + 1$." Between time interval, for which solution is selected, and time interval with reference information (Figure 3) there is a log equal to one decade, as actual state of the system over time interval "t" is not known yet. Thus, the question contains basic uncertainty at the present time at *Operational Planning* stage:

> *How are all types of information available in interval "t" used in finding solution for time interval "t + 1," at Operational Planning stage?*

Only correct algorithmization of these rules for using available information will enable to uniquely determine criterion (or criteria) of water distribution, based on which necessary content of indicators will be formed. Of course, rules themselves can be different for years with different water availability, but the main thing is that they should be known before the beginning of vegetation; this is the second, no less important reason for clear algorithmization of these rules, their *scoring*, in order that after making up a water use plan all participants know on what principle the further water allocation will be implemented, and, thereafter, what measures they (participants) can take to improve their economic activities. In proposals of our colleagues from IWMI, it is recommended to transfer the emphases from *planning management* to *demand management*. These recommendations make sense only for *Operational Planning* stage, because exactly *demand management* is exercised at *Annual Planning* stage, since requirements by participants are *demand for water*. At *Operational control* stage, only daily implementation of found solution takes place, in the context of a real time and actual flow hydrograph, i.e., the issues of correction are not considered. At *Operational Planning* stage, all information on actual *demand* is contained in *request* for volume of water resources supplied to participants, at time point "t" for time interval "$t + 1$." Let's consider possible options of events development in the system while it reacts to demand. Suppose that all participants are sufficiently qualified, and that they properly defined irrigation terms and depths for all farms, taking into account actual development of agricultural crops and forecast on weather conditions, for time interval "$t + 1$." Undoubtedly, such *requests* will more adequately reflect volumes of water resources necessary for concrete participant, but only on indispensable condition that *requests* be completely executed. Under the condition of water deficiency (when *requests* are incompletely executed), they will begin growing gradually, i.e., more informed participant will make too high request for value of supposed curtailing that, in its turn, will even greater increase deficit in the system; such practice took place as early as Soviet times. But this is not a single drawback of direct reaction to demand. As everybody knows, there is a high correlation between precipitation and surface flow in the Ferghana Valley, therefore actual flow hydrograph and actual requests of participants will be in opposite phase, so the probability of coincidence between maximum water volumes claimed by participants and minimum flow values in the system rapidly rises. Such a situation arose as early as summer 2003 at Aravan–Akbura Canal where regardless of high water availability in 2003, one failed to cancel the deficit, really by another reason, by the terms of ultimate system capacity[2]. This analysis shows *that rejection of reaction to planned values will increase deficit depth in water resources allocation rather than reduce it.*

The next small example is not related to attempts to regulate the depth of water resources deficit in the system, it significantly reflects water distribution equitability conditions. Let's consider two hypothetical participants, one of which tries to follow planned values of water volumes allocated to him, through appropriate shifting of sowing terms and first irrigation dates, for diverse agricultural crops; the second one, call him "free artist," orientates exceptionally on weather conditions and crop requirements. Suppose that both the participants are sufficiently qualified to carry out this activity and equally participate in implementing planned indicators established by the government. In order not to complicate analysis with excessive details, let's accept that the year under consideration has mean water availability characteristics, so actual values of flow volumes approximately correspond to crop water consumption norms. Planned values of requirements from each participant and probable flow hydrograph for three decades of vegetation period are given in Figure 5.

In Figure 6, actual values of requests from participants and flow hydrograph are given, at that it is stipulated that when water resources are available, requests are completely fulfilled. Water distribution in decades "$t - 1$" and "t" is trivial, and does not demand explanations; everyone received water as much as he asked, at that excess flow released to river (specific conditions of the Ferghana Valley, which can not be fulfilled in other regions). In decade "$t + 1$," a deficit stipulated only by the request of "free artist" is observed, the

[2] Analysis of the situation with deficit at Aravan–Akbura canal in vegetation in 2003, made by manager of activity 1 N.N. Mirzayev.

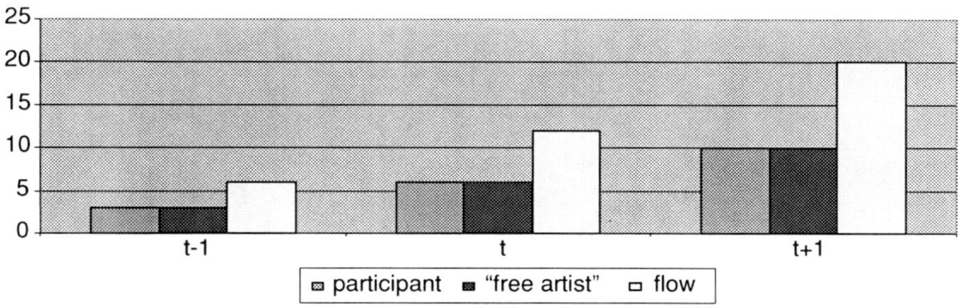

Figure 5. Planned values of requirements from participants and flow hydrograph.

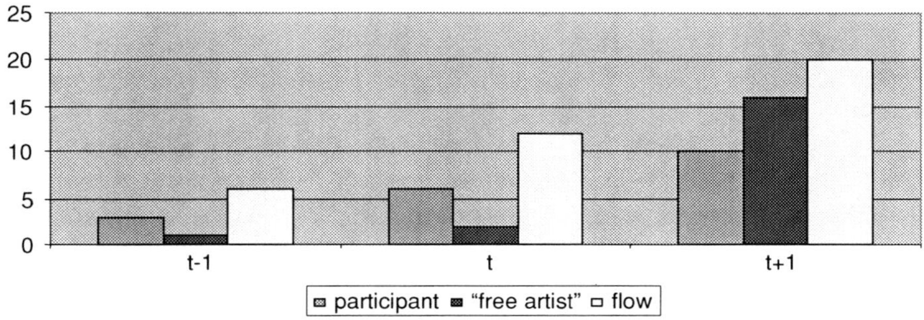

Figure 6. Actual requirements of participants and flow hydrograph.

question is how deficit in the system should be ***fairly*** allocated? The version of proportional curtailing by demand is given in Figure 7. As seen from Figure 7, both the participants received less water, at that "free artist" received 12.3 conventional units of water, while the first participant received 7.7 units.

If suppose that the first participant would adhere to the strategy of "free artist," then deficit depth will rise two times; in the first and second cases the deficit amounted to, respectively, 6 and 12 units. Requests and allocation of

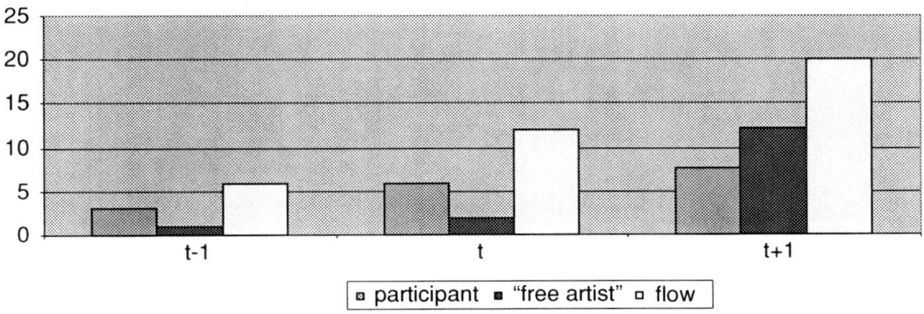

Figure 7. Version of proportional curtailing by demand.

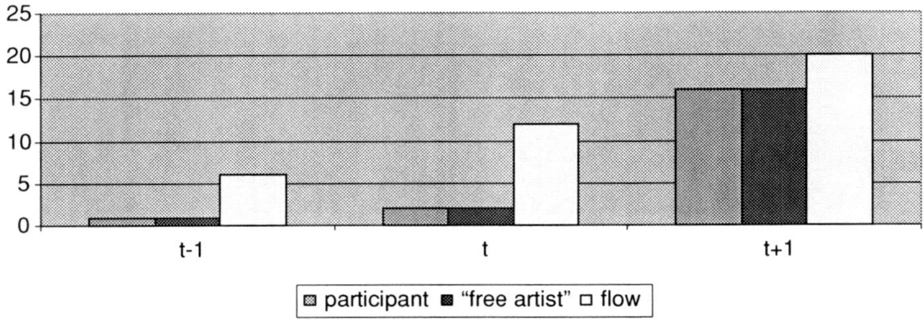

Figure 8. Requests and flow (second case).

deficit according to proportional curtailing by demand are given in Figures 8 and 9.

As seen from these figures, under the conditions of deficit in mutual "freedom," each of both the participants receives 10 units of water resources, therefore achievement in deficit reduction in the first case is identified only by actions of participant following the plan. However, in proportional curtailing by demand, both were punished, at that in the first case the "free artists" received water more by 23% than in the second case; *is that an equitable solution?*, because these 23% are received due to curtailing the first participant who managed to organize his agricultural production taking real restrictions on irrigation system into account that is not small (but may be significant) art in comparing with a strategy mainly oriented to weather conditions.

These examples show that water resources distribution rule (or rules) should differentially consider requests of different participants according to their background and real contribution to ensuring sustainable performance of the system as a whole. Difficulties of formulating these rules are related not only to uncertainty of relations between participants, but also to system

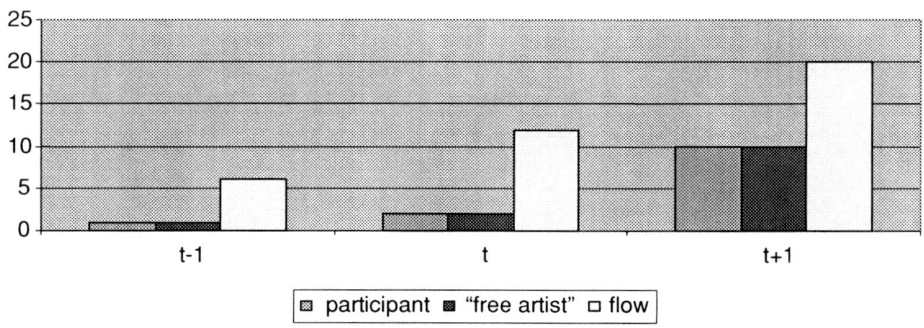

Figure 9. Allocation by requests and flow (second case).

dynamism when actual impacts become apparent only after some time (dead time). Thus, at *Operational Planning* stage, water distribution assessment criterion should include:

(A) information on participant:
- planned values of water supply for period $\{0 \,..\, t+1\}$
- claimed values of water supply for period $\{0 \,..\, t+1\}$
- actual volumes of water supply for period $\{0 \,..\, t-1\}$

(B) information on system:
- planned values of water supply to system for period $\{0 \,..\, t+1\}$
- actual volumes of water supply to system for period $\{0 \,..\, t-1\}$
- actual values of water supply to all participants for period $\{0 \,..\, t-1\}$
- predicted values of inflow to system for time interval "$t+1$."

Based on dataware of points A and B, a set of indicators necessary for *Operational Planning* stage is formed. Here, term "necessary" is knowingly used, since an "adequate" set of indicators will be identified only after analysis of *Operational Planning* stage when *system controllability* parameters are determined.

4.3. OPERATIONAL CONTROL

The main complexity of *Operational Control* is in coordinating work of a system of hydraulic structures with dynamics of water movement in irrigation system. Its great importance is that feedback information on actually implementing established values of water distribution between participants begins to come just from it, and just it transforms actual hydrograph into required form. At *Operational Control* stage, as a rule, any economic criteria are absent, and the main performance indicator is rate of fulfilling established values of water intake for decade and range of fluctuations in water flows at transect of each participant for separate time points. *Operational Control* is carried out by using automatic regulation devices tuned upstream or downstream, depending on concrete conditions and used regulator. Range of used flow regulation facilities is very wide, beginning from the simplest gate and ending with control station equipped with modern computer techniques. As a rule, the higher technical equipment of control station is the less the time of reaction to changes in system is, therefore, the range of fluctuations in instantaneous flow values is. The latter indicator indirectly reflects requirements to system controllability, moreover, it raises accuracy of supplied flow accounting, and therefore provides best control of water distribution. Besides the indicators reflecting water distribution between participants, a set of indicators relating to parameters of water flow in irrigation system is used at *Operational Control* stage. This group of indicators relies on values of levels and flows on canal stretches, because normal performance of any water intake structure is possible only if level of water in diversion canal is not lower than some minimum providing opportunity of water intake. This minimum level id dictated by the construction of concrete water intake facility, and this level is kept by supplying high flow values over the main canal, or other hydraulic structure installed directly at canal (control structure) using the so-called "backwater slope." The task to match operation modes of control structures and intake structures of participants is called "task to stabilize regime of water distribution in the system." This task is central for *Operational Control* stage, as the concept of "*system controllability*" is concretized just by solving it. Specific methods for solving this task as well as factors dominating in control are different for diverse hierarchy levels and for diverse types of irrigation systems, but generalized criterion for assessment of "*system controllability*" can be sufficiently simply formulated, if the concept of "*control efficiency*" is used. The latter is defined as the relationship below:

$$\text{“control efficiency”} = \frac{\text{“summary water flow required by participants”}}{\text{“water flow at the system head”} \times \text{“technical efficiency.”}} \qquad (4.1)$$

Therefore, generalized criteria for assessment of "*irrigation system controllability*" can be defined as "*value inversely proportional to additional water flow to be supplied to the system to meet requirements of participants.*" Note that in real irrigation systems, requirements of participants are almost never fulfilled exactly, that is why assessment of controllability is made through indirect indicators:

- water availability at the system head compared to average weighted water availability for all system participants;
- unevenness of water supply to system participants;
- unevenness of daily water supply to concrete participant;
- relative unevenness of water supply to participants as moving away from the system head.

Each indicator reflects certain aspects in *system controllability*, and enables to focus attention on certain direction of control process as well as on separate sections of the system with the largest deviations by any indicator. For example, availability of water for participants usually declines as moving away from the head. This decline in water availability may be caused by two reasons: excess water intake by upstream participants, and improper distribution of losses along the canal. The first reason is mostly related to poor accuracy of accounting water volumes taken by first upstream participants, and the second one is related to strong nonlinearity of function of losses along the way, that is why attempts to solve this task by using constant (mean or maximum) efficiency values are certainly unworkable. This problem results in initiating a range of methods somehow related to the concept of "water rotation." One of the simplest methods is based on the rule that "water distribution begins from providing the last participant (remotest from the head)." Here the method for water distribution is as follows: based on available flow (or canal capacity if water is not limited), on the principle of maximum remoteness from the head, group of participants is formed. The participants are included in group one after another, as long as available flow will not be depleted, then water distribution for this group is carried out. On the same principle, the next group is formed, and the next step in water distribution is made and so on, until all participants are provided with water. This approach gave a good account of itself on canals with tree structure (one head, no loops) and equal requirements from participants. The range of real conditions for applying it is too limited. Especially, significant problems in using it arise when water supplied over several canals, in this situation the last participant may become first and vice versa, depending on what canal water is supplied over. Under real conditions, another method for "water rotation" is widely applied, the essence of which is as follows: total control time interval (e.g., decade) is divided into sequential time subintervals (e.g., a day, two days, etc.), and all participants are divided into groups by simultaneity. Thus, water is distributed only between participants of respective group in each time subinterval, at that if time subinterval is rigidly established, then group participants vary, and if members of groups are rigidly established, then time subintervals vary. The less time subinterval is selected, the smaller the number of participants is in respective group as well as the more accurate water distribution is. The flexibility of the second method for "water rotation" is much higher, but putting it into practice is significantly difficult and requires application of more advanced methods for calculating water distribution.

The last task of *Operational Control* stage is to account and control allocated water resources, which forms feedback loop at *Planning* stage. Correct solution of this task is related to availability of gauging stations located properly, organization of information gathering in required time intervals and with accuracy of gauging facilities. Thus, a set of basic indicators in task of accounting and controlling should reflect:

- location of gauging station
- time intervals between measurements
- characteristics of gauging equipment.

More frequent mistake made in placing gauging stations in irrigation systems is related to application of rules used for electric circuits like "the quantity of measuring devices equals the quantity of incoming elements plus quantity of outgoing elements minus one." The reason is water losses, which can not be identified by analogues, owing to individuality of each hydraulic structure, so the number of gauging stations should strictly correspond to the number of distributive object elements. Besides gauging stations that ensure accounting of water distribution between participants, a need for periodically setting transient gauging stations to identify parameters characterizing water evaporation and seepage losses arises on long canals. Transient gauging stations are set on sites that have abrupt change in canal parameters, and mainly used in passing maximum and minimum water flows. After constructing a stable curved relationship between flows and losses, a need for transient gauging stations usually disappears. In the existing water distribution system, measurements of water flows are carried out from one to three times a day on KCK and $AB\Pi$ objects, and to 24 daily measurements in pilot fields of *Private Farm* objects. The frequency of measurements in pilot fields should be considered as investigated one, required by conditions of physical–chemical object parameters identification rather than dynamics of water distribution. So, let's orientate to about 8–12 hour time interval of measurements, excepting objects (KCK level) equipped with automated systems. More precise estimation of time intervals and requirements to measurements and applied gauging equipment can be made after performing tasks on "analysis of the current water distribution system," activities 1 and 2 of this project.

5. Water Distribution Control Criteria

5.1. CONTROL PROBLEM SETTING

Applying control theory methods to analyze water resources distribution requires preliminary formalization of the entire process being studied, from irrigation system structure to real control actions carried out by people participating in this process. The result of this formalization is mathematical model reflecting to some extent various aspects of real processes arising in irrigation systems. According to the established language canonization in modern control theory [6], we consider irrigation system as a dynamic object described by the system of ordinary differential equations:

$$\frac{dw}{dt} = g(w, u, \xi, a, t); \tag{5.1}$$

at that: w—n-dimensional phase vector, u—k-dimensional ($k \leq n$) vector-function of control, ξ—m-dimensional ($m \leq n$) vector of external influences,

a—l-dimensional vector of design and phenomenological object parameters, t—time.

For irrigation systems, vector $w(t)$ represents water volumes in diverse regions of space occupied by irrigation system, its values are called phase coordinates, and its dynamics—phase trajectory of the system. Vector $\xi(t)$ of external influences usually represents hydrographs of flow, which is formed in diverse regions of space, and influences irrigation system in a random way. This vector is given through respective distribution functions. Vector-function of control $u(t)$ corresponds to hydraulic structures flows inside and outside the system, which are selected by some person (or persons), for achieving set targets. Value selection capability $u(t)$ are usually limited by both hydraulic structure parameters and water availability, therefore:

$$u(t) \in U(w, a, t); \quad \forall t \in T; \tag{5.2}$$

where: T—control period.

Furthermore, phase variables themselves are limited by construction of irrigation system and adopted operational conditions.

$$w(t) \in W(a, t); \quad \forall t \in T; \tag{5.3}$$

The system of Equations (1) with limitations (2) and (3) is called controllable system, at that expressions (2) and (3) are respectively called possible control space and phase limitations. Since controllable systems are created to achieve something, the main task of control is to define function $u(\bullet)$ by given *control objective*. Usually, *control objective* may be formulated as maximization or minimization of some functional from control, but, as mentioned above, some objectives are pursued in irrigation system control, therefore control search $u(\bullet)$ should be carried out in such a way that several functionals at once reach extreme values. Such mathematical problem, under real conditions, are usually unsolvable, because lots of admissible control providing diverse functionals with extreme values are found non-crossing, for instance, characteristic "availability of water for irrigated areas and water consumers" demands full admission of irrigation system, and characteristic "volumes of water losses in the system" is quite the contrary, i.e., their extreme values are not compatible. The methods of control theory enable to avoid such difficulties. For that, a principal control objective is first formulated in the form of functional that corresponds to the main purpose of irrigation system in the best way, namely: *water coming into the system should be delivered to given regions of space at necessary time points*. Formally, requirements to the system concerning water delivery and actually supplied flows are expressed by equidimensional vectors $q^*(t)$ and $q(t, u)$—j-dimensional ($j \leq n$), and *objective functional* is marked through $\aleph^0(u(\bullet))$. Marking *objective functional* with index "0" shows[3] that in the process of searching controls there will be other functionals in addition to this functional, but for all the next functionals, control is searched only in the area remaining free after finding extreme values of the previous ones. The second functional $\aleph^1(u(\bullet))$, having a special name as well, is *control quality functional*, which reflects the cost or achieving the main objective on condition that the method for achieving it is not single. Formally, limitations (2) and (3) can be reformulated in the form of various functionals, but we will not do that, because the given form is more natural for engineering evaluations. For irrigation systems, *objective functional* can be formalized not uniquely, each of its form will reflect the present understanding of researcher or decision maker what kind of distribution of water coming into the system is the best. So, let's consider several expressions for *objective functional*, leaving final selection for discussion by all project participants. Common expression for *objective functional* can be written in the form below:

$$\aleph^0(u(\bullet)) = \int_{t \in T} \varphi(q^*(t), q(t, u))dt \to \max_{u \in U}. \tag{5.4}$$

Therefore, selection of function φ contains diverse options for determining the best water distribution. To complete formulating this task of irrigation systems control in differential form, it remains only to indicate, among what functions extreme value (***sup***) will be searched. Task (1)–(4) relates to the class of optimum control tasks with fixed time, for which feasible solution is totality $(w(t), u(t))$ in fulfilling the following requirements:

1. vector-function $u(\bullet)$ is defined and piece-wise unbroken for time interval $\{T\}$;
2. condition (2) and (3) is fulfilled for all $t \in \{T\}$;
3. functions $w(t)$ are differentiable at all points, except points where $u(\bullet)$ discontinues, (1) are fulfilled at all points of differentiability;
4. vector-functions $q^*(t)$ and $\xi(t)$ are defined and piece-wise unbroken in time interval $\{T\}$, (these functions are uncontrollable, since the first one characterizes requirements by irrigated areas and transit, and the second one characterizes flow hydrograph).

Extreme values in task (1)–(4) will be searched just among such feasible solutions.

Depending on detailing Equations (1)–(3) and criterion of control (4), a chain of tasks is formed, of which the first one is called task of planning water allocation in irrigation systems, based on its results volumes of water

[3] Here and further, upper indices signify belonging of variables to any semantic group, and lower indices are used as counting ones, on which summation and sorting are carried out.

resources supply are specified in accordance with established limits, and operation mode of waterworks is planned. In terms of optimum control theory, this task is classified as a task of identifying program trajectory of the system. The next task is task of correcting, which in terms of optimum control theory is classified as a task of control synthesis or projecting of feedback statement. This task [1] fundamentally differs from task of identifying program trajectory in both quality criteria participating in forming functional and solution methods. This task is set to implement found program trajectory, which is final in any process of control. The importance of solving this task is defined by that if mistakes or deviations in defining program trajectory lead to economic losses (achieving objective is more costly, or values of some indicators are rather worse than it is proposed), then poorly projected feedback statement may completely destroy control system, for instance, dam break formation on separate canals abruptly reduces controllability of the whole irrigation system that results in absolutely incommensurable water losses. In general, the task of synthesis is more difficult than usual task of optimum control due to the lack of regular methods for solving it as well as lack of necessary conditions, which serve as an initial point in constructing design models, but here advantages of GAMS package, which frees researcher from the necessity of searching a method for solving and enables to concentrate on correctness of formulated mathematical model and adequacy of criteria for water allocation, becomes apparent in full.

5.2. CONTROL CRITERIA IN ANNUAL PLANNING

Formulation of main control criterion of *purpose functional* will be based on basic operational performances of irrigation system functioning, see Section 2. First, we should adopt conditions of all participants independence, it means that water under- or over-diversion by specific participant is reflected only on himself, in this case function φ will be additive.

$$\varphi(q^*(t), q(t, u)) = \sum_{j \in J} \lambda_j \times \psi(q_j^*(t), q_j(t, u)). \tag{5.5}$$

Conditions of all participants' independence is very important assumption in our analysis, since namely it allowed expression (5) in form of sum, if there is water exchange opportunity between participants ignoring system, expression (5) will be wrong. In this case, it is necessary to build oriented graph of possible water exchange between participants, and summarize fact deviation from requirements already upon it. This water allocation option is used under modeling of *private farm*. It is required to clarify record meaning $q(t, u)$, since in optimal control theory the record $q(t,u(\bullet)$ is used more often to underline variable dependency on control function, their equivalence $q(t, u) \equiv q(t,u(\bullet))$

will be assumed in given paper, if variable dynamics from control is considered in specific point of time, designation $q(t, u(t))$ will be used. Expression (5) has not determined yet, and to complete setting (to formula that assume numerical implementation) metrics is to be introduced that will provide calculation and comparison of deviation between demanded and actual flow for each participant. Just selection of metrics will define water allocation rules between participants, therefore we will consider several options. The simplest metrics is root-men-square, i.e.,

$$\psi_j(t) = \text{-}[q_j^*(t) - q_j(t, u)]^2; \quad \lambda_j = 1; \quad \forall j \in J, t \in T. \tag{5.6}$$

Symbol "-" in expression (6) means that quadratic deviation minimizing in water supply over all system will be fulfilled in (4). Main disadvantage of expression (6) is infringement of participants with small water diversion amounts. That's why in practice [7], usually normalization of deviations through total water supply amount of participant is used, namely:

$$\psi_j(t) = \text{-}(q_j^*(t) - q_j(t, u))^2; \quad \lambda_j = 1/w_j^{*2}; \quad \forall j \in J, t \in T; \tag{5.7}$$

here:

$$w_j^* = \int_{t \in T} q_j^*(t)dt; \quad \forall j \in J; \tag{5.8}$$

Instead of quadratic deviation, deviations on module can be used in expression (7), however, it is rather matter of taste, than real improvement of water allocation. More crucial water allocation improvement can be achieved, if consider water shortage and excess as unequal, for example:

$$\psi_j(t) = \text{-}\alpha_j(t) \times (q_j^*(t) - q_j(t, u))^2; \; \lambda_j = 1/w_j^{*2}; \; \forall j \in J, t \in T; \tag{5.9}$$

here:

$$\alpha_j(t) = \{\alpha_1 \; npu \; q_j^*(t) > q_j(t, u); \quad \alpha_2 \; npu \; q_j^*(t) < q_j(t, u)\}. \tag{5.10}$$

At $\alpha_1 > \alpha_2$, water under-supply effects will reduce in system, and at $\alpha_1 < \alpha_2$, on the contrary, over-supply effects.

Interesting options in water resources allocation emerge, if guarantee conditions[4] are used, in this case water supply of participant $q_j^*(t)$ should be

[4] V.A. Dukhovny proposed to consider: if in expression (11) to divide over hierarchies components' management, management $q_j^{\wedge}(t)$—to keep with participant, and selection $\Delta q_j^*(t)$—to charge to the system (all participants), multicriterial task will be obtained for analyzing conflict situations that will belong to the cybernetic systems category studied by, Y.B. Germeyer [1]. Important peculiarity of these systems is that it is possible to formalize homeostasis notion as well as identify sustainable equilibrium areas (conditions of collective equilibrium for all participants).

presented like:

$$q_j^*(t) = q_j^\wedge(t) + \Delta q_j^*(t), \tag{5.11}$$

where the first item is guaranteed water supply of the participant, and the second one is addition required for its farming activity. Naturally, target function should respond in a special manner for guarantee conditions. Below target function option ψ meeting these conditions is given:

$$\psi_j(t) = -[\Delta q_j^*(t)/(q_j(t,u) - q_j^\wedge(t))] \times (q_j^*(t) - q_j(t,u))^2;$$
$$\lambda_j = 1/w_j^{*2}; \forall j \in J, t \in T; \tag{5.12}$$

Expression (12) allows keeping water supply values that are not lower than quarantined, however issue of determining guaranteed water supply values themselves remains open. As one option of this task solution, investigation of support task with flow hydrograph with very high water availability $\xi(t) \sim 80 \div 90\%$ and purpose criteria (7) can be suggested. As a result of this task solution, values $q(t,u)$, which can be adopted as $q^\wedge(t)$, will be obtained, and then come back to solution of basic task with target criteria (11), (12), and flow hydrograph $\xi(t)$, established for considered year. For irrigation systems, this approach is possible due to that with reducing flow water availability percentage (actual water amount increase) permissible control space expands, that is $U(w, a, t)|_{\xi(t)>80\%} \subset U(w, a, t)|_{\xi(t)<80\%}$. In addition to direct flow values $q^*(t)$, their price equivalents determined through productivity of one cubic meter of water, or one irrigated hectare for each participant, can be applied in expressions (7) and (12). However, introduction of economic performances directly into *target functional* should be carried out only after thorough analysis of participants' farming activity, since differences in water productivity leading to *inequality* of participants may be caused by different reasons not depending on them (participants), for example, soil parameters under agricultural lands, irrigation network efficiency, etc. At given stage of the project implementation it is more expedient to integrate all water losses components in irrigation system in own functional (*control quality functional*), which will be minimized following basic functional (*target functional*). Expression for *control quality functional* on water conservation can be presented like:

$$\aleph^1(u(\bullet)) = \int_{t \in T} (\eta^T \bullet w(t))dt \to \min_{u \in U}{}^5; \tag{5.13}$$

[5]Expression $(\eta^T \bullet w)$ presents inner product of vectors defined as:

$$(\eta^T \bullet w) = \sum_{j \in n} \eta_j^T \times w_j,\ n- \text{ vectors length } \eta^T \text{ and } w.$$

where: η^T-technical canal site efficiency defined as ratio of outgoing flow to incoming one under lack of water consumption on this site on part of participants, vector, which length is equal to vector length w. Water resources saving criteria simplicity is related to "lucky" option of phase variables for describing irrigation system dynamics.

5.3. CONTROL CRITERIA IN OPERATIONAL PLANNING

The next stage of water allocation optimal control problem solution is planned flows correction stage in accordance with real existing water situation. To this end, total control time period {T} is separated on three non-crossing intervals $\{t^0 : t - 1\}$-past, $\{t\}$ present, and $\{t + 1 : \mathrm{T}\}$ future. All variables related to $\{t^0 : t - 1\}$ are known, they reflect real water supply and allocation schedule that will differ from optimal one, obtained by calculating on planning stage, on three reasons:

- deviations in water resources allocation among participants caused by limited abilities of hydraulic structures,
- deviations in actual water supply amounts $\xi(t)$ to irrigation system,
- deviations in requirements $q^*(t)$ of water resources by participants caused by different climatic factors.

Totality of these three reasons is forming in-system balance for period $\{t^0 : t - 1\}$, which makes water flows correction necessary for next time periods. Before formulate functional options for water allocation, it worth to note that additional vector appears *at operational stage*, let's designate it through $q^{\sim}(t)$, length "n," which reflects applications of participants based on real climatic conditions and economic activity of the participants themselves. The feature of this vector is that in contrast with its analogue $q(t)$, therefore, currently water supply values to participants are corrected only for time period "$t + 1$." Field surveys results show big difference between values of applications and requirements, in particular in the beginning of vegetation period, however, their integral parameters (total water diversion amount) become closer with increasing period $\{t^0 : t\}$.

Specific feature of agricultural production is that basic diversity in water demand between different participants is defined by diversity in agricultural crop sowing and first irrigation dates. Hence, once first irrigation dates expire, it is necessary to amplify vector $q(t)$, using data included in applications and actual flow hydrograph during past period of time $\{t^0 : t\}$. Presently similar amplification is absent, therefore, in fact, participants' applications are single document that determine water supply amount amplification for participants for the next decade. In Section 4.2 one example is given showing how unconditional system response for applications can cause various options of different participants' interest infringement. To amplify vector $q(t)$ we will form new requirements $q^*(t)$ according to formulae:

$$q^*(t) = \begin{cases} q^*(t) = q^R(t) \forall t \in \{t^0 : t\} \\ q^*(t) = q(t) + \delta q(t) \forall t \in \{t + 1 : T\} \end{cases}; \tag{5.14}$$

where: $q^R(t)$—actual water supply to participants, and $\delta q(t)$—balance discrepancy for expired period of time, is calculated like:

$$\delta q(t) = \frac{\int_{t^0}^{t} (q^R(t) - q^{`}(t))\, dt}{T - t}. \tag{5.15}$$

Using new vector of demands $q^*(t)$, for the rest part of period $\{t + 1 : \mathrm{T}\}$, fulfill solution of task (1)–(4) at criteria of type (7). Obtained vector $q(t)$ will be necessary water use plan amplification. Further water supply amount amplification to participants can be calculated by formula:

$$q^R(t + 1) = (q(t + 1) + q^{\sim}(t + 1) + \delta q(t))/2, \tag{5.16}$$

where balance discrepancy $\delta q(t)$, as before is calculated by Formula (15).

5.4. WATER ALLOCATION CRITERIA IN OPERATIONAL CONTROL

Operational control provides solutions implementation on water resources allocation developed on previous stages (annual and operational control) by adjusting hydraulic structure parameters to supply established water flow.

6. Conclusions

1. Irrigation system control task belongs to multi-criteria tasks, where target uncertainty requires introduction of additional assumptions about significance of one or another parameter, without which it is impossible to complete criteria convolution.
2. Each irrigation system control stage demands own set of parameters and criteria providing adequate assessment of solutions (control actions) adopted on specific stage.
3. The most important irrigation system control task is correct determination of target functional that provides both cost-effectiveness of water allocation process and system sustainability in control under various deviations of participants from formulated water allocation rules.

Bibliography

1. Moiseyev, N. N. (1981), *Mathematical Tasks of System Analysis*, Moscow Nauka, p. 487.
2. Peshel, M. (1980), *Modeling of Signals and Systems*, Paris, Mir, p. 300.
3. Dukhovny, V. A. (1983), *Irrigation Complexes on New Lands of Middle Asia*, Tashkent, Uzbekistan, p. 184.
4. Lauks, D., Stedinjer J., and Heit, D. (1984), *Planning and Analysis of Water Supply Systems*, M., Energopromizdat, p. 400.
5. Sage, E. P., White, Ch. S. (1982), III "Optimal systems control", M., «Radio i svyaz», 391 pp.
6. Pontryagin, L. S., Boltyansky, V. G., Gamkrelidze, R. V., Mishchenko, E. F. (1983), *Mathematical Theory of Optimal Processes*, Moscow Nauka, p. 392.
7. NATO SFP 974357 (2000), "Integrated Water Resources Management for Wetlands Restoration in the Aral Sea Basin".
8. Dukhovny, V. A., Tuchin, A. I. (2001), «Irrigation systems control», Coll. IFAS, SIC ICWC №5, Tashkent, pp. 4–10.
9. Umarov U., Tuchin, A. (1986), Operational water sector control based on automated systems.—In coll. Water resources development issues in arid zones. UN Economic Commission for Africa, Ministry of Reclamation and Water Management USSR, SANIIRI. Tashkent, pp. 300–314.

THE FUTURE OF THE PRIARALIE

T. Kamalov
Chairman of State Inspection "Gosvodkhoznadzor"

1.1. SENIOR EXPERT OF STATE INSPECTION—SH. TALIPOV

Due to Aral Sea area reduction and a volume decrease of one-sixth in just 40 years, we have become eyewitnesses to a real ecological crisis, which has impacted upon millions of people living in this region and beyond. The Aral Sea basin has ceased to be an enormous natural conditioner that kept back cold airflow from the north and cooled air from south. Over the last years, the desertification process has covered more than 4 million hectares of land, whilst landscapes adjacent to the Amudarya and Syrdarya River deltas have been strongly impacted.

The main natural sources of pollution in the atmosphere are the deserts of Karakum and Kyzylkum, as well as the dried Aral Seabed. Between 75 and 125 million tons of salts are removed annually from the Aral Sea basin. This accumulates on an area of between 1.5 and 2.0 million km^2, causing great damage to agrarian regions located nearby by poisoning the atmosphere, soil, lakes and rivers and, eventually, living organisms. The majority of the population living in irrigated regions of Central Asia consume water from irrigation canals containing various salts, fertilizer residues, pesticides, nitrates, and other wastes. Poor quality and a lack of drinking water is a cause of many diseases.

Ecological problems in the Aral Sea have caused a worsening in the local population's health as a result of water and air pollution, lack of water supply sources, and poor sanitation. Degradation of flora and fauna in Priaralie is an especially critical problem that negatively impacts on the economy, and therefore on living standards in neighboring regions. In the epicenter of ecological disaster areas, indicators of the rates of infant mortality and death during childbirth are the highest in the CIS area. Here, middle life expectancy is reduced and tuberculosis, anemia, thyroid gland dysfunction, kidney and liver diseases are widespread. Blood diseases, cancer, asthma, and coronary deficiency are also progressing. Traces of pesticide were found in women's milk. All this leads to population degradation.

A critical situation can also be observed in terms of population employment, in particular in economic sectors related to agricultural production

P. Wouters et al. (eds.), Implementing Integrated Water Resources Management in Central Asia, 81–86.

where, due to irrigated land degradation, production volume has declined alongside a labor efficiency decrease, leading to unemployment growth and population impoverishment.

The "*International Fund for Aral Sea Saving*" (IFAS) was established by the heads of state of Central Asia on March 26, 1993 in Kzylorda as a basic tool for solving the problems in the environmental and social-economic situation in the Aral Sea basin. Due to intensive activity on the part of the heads of state of Central Asia, a set of important documents were adopted within the IFAS framework:

- Nukus Declaration—with responsibilities for cooperation at regional level based on mutual respect, good neighborly relations, and a determination to overcome the ecological crisis in the Aral Sea basin, as well as its negative impact on man and nature; the declaration called on the international community, governments, and nations around the world to help in realizing joint efforts.
- Almaty Declaration—called on the United Nations Organisation to pay special attention to the existing critical situation in the Aral Sea basin and to undertake powerful measures on environmental protection, and to offer assistance to the suffering region and its population; the declaration confirmed a readiness to support international organizations and institutes in their activities regarding the implementation of a Programme of Concrete Actions and other regional projects.
- Ashkhabad Declaration—recognized the importance of integrated problems–solution related to the enhancement of the social-economic situation in the Aral Sea basin; called for the strengthening of activity aimed at securing the involvement of international organizations in the implementation of the Aral Sea basin programs and projects; provided help and support in the implementation of the project "*Aral Sea Basin Water and Environment Management*;" implemented a set of measures and projects of priority-driven aspects of Aral basin population protection; enhanced, through training and other programs, population knowledge, and understanding of issues of environmental conservation, effective water use, and habitat improvement for present and future generations.
- Dushanbe Declaration—defined priority measures on social-environmental conditions improvement, aimed at providing the Aral Sea basin population with clean drinking water; speeded-up the development of practical measures on collector-drainage water use amount increase; recognized necessity of creating special UN Commission responsible for coordination of international organization's and country-donor's activity; improved monitoring system and awareness among countries of the water situation, as well as that

of other natural resources, so as to be better placed to make decisions on their effective utilization; called for the implementing of results obtained during the development of water and power management projects for effective and mutually beneficial cooperation.

These, and other documents reflecting the situation, which threatens not only Central-Asian states but also all humankind, were widely appreciated and supported by the world community, including the governments of many countries as well as a wide range of international institutes. With the increasing scale of the Aral ecological disaster, the world community better understands that such crises can only be overcome through joint and coordinated efforts.

At the same time, implementation analysis of the "*Program of concrete actions on environmental situation in the Aral Sea basin for the next 3–5 years taking into account social-economic development of region*", adopted in 1994 in Nukus, points towards the main factors leading to a failure to implement sets of programs and projects in full volume being a lack of financing and poor coordination and project management, both on the part of IFAS executive bodies and beneficiaries.

In spite of a relative reduction in water diversion amounts, Aral Sea and Priaralie lakes deterioration continues because of the imperfect system and uncoordinated actions of the various parties involved. Currently, the Aral Sea water level is 31.0 m BC, indicating a decline of 22 m. over the last 30–35 years. It should be determined at what level, and under what conditions, and by means of what sources of water resources, the sea should be maintained. Based on these decisions, documents could be produced in order to adopt an official plan regarding the future of the Aral Sea that will allow the development of an integral program of Priaralie ecosystem rehabilitation.

Up until 2001, due to inadequate management of the Aral Sea Coastal zone, conflicts of interest, non-clad structures and dams, and insufficient supply of mechanisms and equipment, a huge amount of water (6.0 km^3) was simply directed, annually, to the Aral Sea and its coastal zone, without any substantial economic benefit (environmental area rehabilitation). Works carried out by the GEF Project "*Water resources and environment management*," under its Component E "*Rehabilitation of wetlands "Sudochie*," like work carried out under its other components, proved to be realistic and accessible action on this stage, which should be further realized. The creation of small, local reservoirs up to the sea shoreline in Amudarya and Syrdarya deltas and adjacent areas of the dried Aral seabed will serve to aid the rehabilitation of the environmental situation in Priaralie, returning the ecosystem to prior levels of achievement, resulting in a productivity increase as well as a reduction in salt and dust transport, and climate aridity. Similar rehabilitation of the river delta water structure will provide fish yield increase, strengthen cattle-breeding forage resources, and create favorable conditions for fur animal breeding, reproduction of waders and waterfowls, and development of forest-tugai bushes. Furthermore, residents of settlements located around Mezhdurechie, along the Kazakhdarya flow channel, and in Mujnak and other canals with water diversion from Mezhdurechie reservoir, will obtain guaranteed conditions for irrigated agriculture development.

However, under the present conditions of limited financial and water resources, this project will be able to provide only partial rehabilitation of the damaged Priaralie ecosystem. To solve the problem of a further area of two hundred and thirty thousand hectares of dried lakes, and the Aral Sea itself, as well as dried sea beds with an area of more than 4 million hectares, the assistance of the world community is needed.

As things stand, parts of the Aral Sea are going to vanish in vast areas of Karakum and Kyzylkum, which are already starting to join together. Since this area emerged from beneath the water, the soil formation process has occurred in combination with processes of water logging, salinization, erosion, and pollution, depending on relief conditions and soils. Regarding this detailed study of ground water, a study of the dried seabed topsoil and the mapping of soil erosion, salinization, and pollution, as well as the mapping of ground water, is required. Based on these results, it would be possible to define basic hydrological, hydrochemical, and toxicological parameters, identify the characteristics of changes in the environmental state of wetted zones during the process of technical measures implementation, evaluate the present state of wetlands' ecosystems in order to rehabilitate existing ecosystems, forecast their development and management, and distinguish areas of the first and further years of their development by tugai forests and other salt-resistant plants.

Maintaining the levels of the Aral Sea at an elevation 31 m BC demands environmental releases, unproductive losses reduction from large main canals, and effective use of all types of drainage, collector, and wastewater, whose amounts reach some dozens of billions of cubic meters. It is worth noting that collector-drainage water amount is not always sustainable, and its quality can cause new problems in reuse locations. An increase in collector-drainage impacts on the agrarian landscapes, leading to agricultural production decline. Therefore, to use wastewater without any risk, integrated assessments of their quantity and quality, formation conditions and regime, and forecasts for a future based on different development policies, is required.

Opportunities for new irrigation development in the region are very limited. In this connection, the focus of requirements in newly irrigated areas development in sustainable economic conditions should be shifted toward the increase of water-land resources use efficiency, and this should be the basis for water use and conservation policy.

In spite of efforts in water resources distribution among water users, even within one country it is difficult to completely avoid water consumption disproportion, in particular between the Amudarya middle and lower reaches. It is necessary to develop effective mechanisms and regulations, taking into account flow losses, oriented toward providing sustainable water distribution, including environmental releases between control sites and irrigation systems, in particular in low water years. The achievement obtained through the joint actions of the Central-Asian states over the last 10 years, where water diversions have been reduced by more than by 7 km^3, should be developed and strengthened.

Regarding issues of water diversion from Amudarya, it is necessary to use water resources within the limits of actual flows, taking into account the maintenance of the existing Aral Sea water surface area, providing environmental releases on Amudarya by reducing ineffective water resources use by means of priority directions.

Efforts made by the Aral Sea basin states based on accumulated international experience in transboundary water resources allocation, and taking into account historical lessons, should be directed toward the creation of the maximum possible cooperation between water institutes, as well as the improvement of joint activity.

Special importance in effective water resources management and guaranteed water supply is attached to the reliability of the technical state and safety operation of reservoirs, water intake, water lifting, and water distribution structures, canals and collectors. Due to the duration of their operational terms, and the financing limitations for their repair and reconstruction, their parameters have changed. Sustainability and reliability have decreased, as a result of which water losses and ineffective water flows have increased.

The need emerged to create a database and modern system for monitoring the technical state and safe operation of both interstate and municipal water objects, adopting general regulations and defining common methods and criteria for their safety, whilst developing a program for their rehabilitation and further maintenance.

In the rural Priaralie area, labor resources are high, and as a result there is a low level of population income. Besides this, difficulties experienced during the transition period as the region adapts to market conditions, in particular under water resources deficit, influence agricultural production efficiency. The creation of pilot-farms equipped with modern production technologies, funded by grants and beneficial credits, would provide concrete support for the population living in the environmental disaster zone, and would initiate business activity among the unemployed of this area.

There is also a great need for schemes such as the transfer of effective irrigation water use skills to residents engaged in irrigated agriculture, the training of experts and managers of agricultural entities in special land use methods directed toward the prevention of unproductive water losses and irrigated land degradation, and the increase of concern and responsibility on the part of public organizations and local selfgovernance bodies, as well as citizens, for environmental region rehabilitation.

Analysis of the present state of, and development in, the Aral Sea basin environment shows further deterioration of the environmental situation in this region. The Aral Sea and Priaralie problem is increasing, and operates like a chain reaction that reveals new problems again and again. This situation demands urgent and immediate attention, and should be the priority up to 2010 and beyond.

Regarding the "*Program of concrete actions on improvement of environmental and social-economic situation in the Aral Sea basin for period 2003–2010*" (ASBP-2), it is necessary to reflect on the environmental situation in Priaralie with reference to the underlying state of the sea itself, and of its coastal areas. It is necessary to identify the negative impact of the ecological crisis on other basin regions up to the water resources formation zone, on irrigated lands, water objects, forests, social life, and the economy, taking into account the main reasons inducing these problems. ASBP-2 projects should be divided into top-priority projects solving problems resulting from the Aral Sea crisis, and accompanying projects should be oriented on the basis of specific area issues in Central-Asian countries, within the framework of a single basin strategy.

It is obvious that it is very problematic to rehabilitate the Aral Sea to prior parameters under present economic conditions. However, rehabilitation of the Priaralie zone, as well as sea state stabilization, is possible. Today, the task is to minimize crisis impact. Therefore, integration of efforts of CAR states, with the underlying support of the world community, is the proper way to overcoming the Aral Sea environmental crisis and save its coastal zone.

SOCIOECONOMIC ASPECTS OF INTEGRATED WATER RESOURCES MANAGEMENT IN CENTRAL ASIA

ON PUBLIC PARTICIPATION IN WATER RESOURCES MANAGEMENT

A. D. Ryabtsev
Chairman of the Committee for Water Resources, The Ministry of Agriculture of the Republic of Kazakhstan, Republic of Kazakhstan

The urgency of the problem of efficient water resources use and protection has been growing with every passing year, in countries across the world, as a result of population growth and increasing industrial and agricultural production volumes, accompanied by the expansion of irrigated agriculture.

Issues of water supply have always been at the center of attention in the countries of Central Asia, since historically the economic development of the region has been in many respects identified with the development of irrigated agriculture and irrigation systems. Prior to the middle of the 20th century, irrigation development rates had been low and the scope of its impact on the environment had not been particularly notable. However, from the 1950s, when the application of powerful construction and earthmoving machinery was introduced and the wide-scale development of new irrigation lands was launched, in the comparatively short period of just 20–30 years the water resources of Central Asian rivers became fully employed in processes of economic use. Furthermore, by the mid-1980's, available water resources from the Syrdarya River had already been overused by 50%. Eventually, such water resources overuse led to negative consequences and the occurrence of the Aral Sea disaster.

In order to overcome these negative consequences, and to improve the environmental and socioeconomic conditions in the Priaralye region and the Amudarya and Syrdarya River basins, the governments of Central Asian countries started taking complex measures, with the cooperation and support of the international community. To develop and coordinate mutually agreed actions aimed at efficient water-energy resources use in the region, a number of scientific-research studies were carried out within various programs and projects, including some which involved engagement with international consultants.

Major works started after the adoption of the "*Program of concrete actions on improving ecological situation in the Aral Sea Basin considering socio-economic development of the region*" at Nukus, on January 11, 1994. After completion of the GEF project "*Water resources and environment*

P. Wouters et al. (eds.), Implementing Integrated Water Resources Management in Central Asia, 89–94.

management in the Aral Sea Basin" it was possible to say that the first stage of tackling the regional problems had been implemented.

During this period of time, vast experience was accumulated, mostly positive, in the sphere of interaction between national agencies and organizations from various sectors of the economy, both at intrastate and interstate levels. The positive experience of interaction should be underlined, in that it has taken shape between state water management bodies of Central Asian countries. It is also important to note the active role played by the "*Interstate Commission for Water Coordination*" (ICWC), which was established during the early years after independence and which proved to be of great significance in establishing water resources management free of any major conflicts or disagreements.

It should be noted that during these years the regional states carried out reforms of all economic sectors concurrently, including water management. Each country chose its own path of development and, in accordance with selected reform concepts, implemented their own agrarian, water resources management, and energy policies. Naturally, the interests of the individual states differed as to their priorities and orientations and sometimes contradicted each other. In these circumstances, the ICWC managed to identify the common ground as regards the mutual interests of the counties concerned, and has, as far as possible, been able to ensure the continued provision to the various sectors of the economy in the basin of an uninterrupted water supply. It is also worth noting that the implementation of the "*Program of concrete actions...*" was carried out with the active support of the ICWC and the various water management agencies and organizations of countries in the Basin.

At the same time, the involvement of the public at large in addressing the Aral Sea problems during this period was minimal. This was determined by a number of factors. Firstly, the countries of the region had only recently obtained independence. Upon independence, they became engaged in the process of identifying models of development, with the result that state institutional structures did not pay adequate attention to establishing civil societies. Secondly, citizens themselves, who were accustomed to living under systems of totalitarian regimes, were not equipped to start actively participating in water resources management processes and addressing water and environmental problems. Finally, there was an absence of trained professionals who could comprehend the overall picture of water/environment processes, and who could tackle these issues at the public level.

Since then, the situation has changed dramatically. A significant number of non-governmental organizations have emerged which address problems in the water sector. These organizations have been giving consideration to water ecosystems and public awareness. They also interact with governmental agencies in the management and regulation of water resources, thus becoming a real force capable of exerting influence on the regulation of water relations at local, basin and regional levels. And so, our time itself calls for broader public participation in order to successfully achieve all the anticipated goals in the field of water/environment management.

Types of public participation in regional water resources management and use may vary, from the involvement of NGOs, movements, and parties pursuing different political goals in the process of improving the ecological situation in the region, right up to broadening mandates for Water Users Associations, Basin Counsels, and other such organisations. Forms of public participation may also vary. Such participation may range from the holding of workshops and conferences, the making of statements in the mass media, and the holding of public hearings, right up to the carrying out of practical actions such as the planting of trees and gardens and the cleaning of water protection zones and strips.

The major goal of such public activities is to draw the attention of governmental agencies and decision makers to the importance of efficient water resources use and protection. In addition, public organisations are indispensable in terms of the environmental knowledge acquisition by civil society members and the younger generation. Only they are capable of attracting everyone's attention, to help instill in the population a cautious and responsible attitude toward water and the environment, as well as in the efficient use of natural resources.

It should be noted with satisfaction that for several years, since the establishment of the Regional Ecological Center of Central Asia (REC CA), much success has been achieved as regards the creation of conditions for uniting efforts undertaken by NGOs in countries of the region, with the aim of improving the ecological situation in the Aral Sea Basin. For example, in 2002, REC CA considered more than 250 proposals submitted by NGOs from all the countries of the region. As a result of competitive selection, project proposals were funded with regard to stimulating public participation in addressing the vital problems of the region.

One such REC CA project is "*Development of recommendations as to practical application of international conventions on transboundary water and energy resources use in Central Asia*." Recommendations have been elaborated within the project on shared water-energy resources use, taking into account adopted international norms and rules. Analysis has been carried out with regard to the implementation of agreements on the Aral Sea river basins as well as the assessment of the applicability of International Law to regional water resources use. Workshops were held in Kyzyl-Orda (Kazakhstan) and Nukus (Karakalpakstan) on the practical application of current provisions of International Law to transboundary watercourses in Central Asia, with the participation of nongovernmental organisations and representatives of regional and local water management organisations. Project team members from Kazakhstan and Uzbekistan have summarized all the materials of the workshop and developed a report in CD-format for wide dissemination.

Representatives of public ecological movements conducted a large-scale ecological action dedicated to tackling problems of environmental protection in the transboundary Irtysh-Ob River system. Participants of the action (26 representatives of public movements and five NGOs from the town of Karaganda) floated on rafts made of plastic bottles along the Irtysh River, following the Ridder–Ust-Kamenogorsk–Semipalatinsk–Pavlodar route. The action lasted for 35 days, and the organizers managed "to stir up" not only the local population, but also both the local and national electronic mass media and press. In the course of this action, the most urgent ecological issues were dwelled upon, among which was the purity of water in the Irtysh and other rivers of the region. Participants of the campaign conducted lectures, discussions, and "round table disputes," accompanied by a specially prepared presentation of an exposition entitled "Banks of My River," which consisted of photographs and children's drawings. Signatures were collected in support of establishing a bio-reservation in the territory of the Kazakhstan Altai. Sets of informational handouts and visual aids were distributed to schools to be used in classes on ecological education. Meetings and sessions were also held involving the local authorities. The authors of the campaign trip gave an overview of the action in the form of a manual entitled "*How to organize an interesting ecological action*."

While implementing the "*Civil society participation in effective natural resources management on transboundary territories in Central Asian region*" project, with financial support provided by REC CA, participants from Kazakhstan and Uzbekistan made an evaluation of water resources use in the transboundary context, and analyzed the national water legislation in the countries concerned. Consideration was given to the possibility of including international conventions provisions in national legislation. Studies were carried out with regard to national programms and the experiences of interaction between national agencies monitoring water use. An assessment was made as to current water use and the role played by the population in water resources management. Participants of the project tried to identify the role of the civil society in both the decision-making process and in ecosystems management. Executors of the project published the information bulletin "Hayot" in the Russian and Uzbek languages, and also a collection of papers entitled "*Water resources and public participation in efficient management (Kazakhstan and Uzbekistan)*." The authors of these publications showed that public participation and involvement in management constitutes a new system of views, involving a coordinated system of interaction between politics, law, regulations, institutions, civil society and consumers.

In May 2003, public organisations of the Central Asian region, with support provided by international donors, held an international conference in Tashkent on the theme of "*Public participation in overcoming water deficit in Central Asia*." Participants of the conference supported the transition of Central Asian countries toward integrated water resources management and expressed their willingness to make NGOs more active in the development of models of sustainable water use which are applied in international, national, and regional water management agencies of Central Asia. The importance of the accession of Central Asian countries to international legal documents (conventions, agreements, and protocols) on transboundary watercourses was underlined, as was the importance of the improvement of national water legislation and its adjustment in order to comply with International Law. Participants considered the involvement of the public in water resources management as the most important direction for the ecological movement in Central Asia. The establishment of a water-oriented NGO network in Central Asia was acknowledged as the next step to be undertaken along these lines. Issues of public participation in programmes and projects of ASBP-2 were scheduled to be further discussed at a special session of the Fresh Water International Forum held in Dushanbe on August 31, 2003.

Representatives of regional NGOs are members of "*Global Water Partnership in Central Asia and the Caucuses*," and take an active role in programms on public participation in the process of Integrated Water Resources Management. To make the public more active in addressing the urgent problems of the region, support from the governmental agencies is of primary importance. A certain amount of work in that regard has been carried out in countries of the region. Thus, the new edition of *Water Code of the Republic of Kazakhstan* (clauses 9, 10 of the Article 9) stipulates that openness and public involvement in addressing issues of water fund use and protection, as well as availability of information about the state of the water fund of the Republic of Kazakhstan, are the principles of water legislation of the Republic of Kazakhstan.

Furthermore, clause 2 of Article 43 of this Code stipulates the establishment of Basin Councils. These are designed to become advisory-deliberative bodies and are to be chaired by the heads of the corresponding basin management organizations. The law stipulates that the Council may include representatives of public associations alongside members of the local representative, executive, and territorial authorities. The Council gives consideration to urgent issues in the field of water fund use and protection, and submits proposals and recommendations for the parties of a basin agreement. In its turn, a basin agreement on water bodies/facilities rehabilitation and protection is concluded between the basin management organization, local executive authorities, and other participants located within the boundaries of the basin, for the purposes of joining and coordinating their activities, as well as implementing measures on water bodies/facilities rehabilitation and protection.

Basin agreements contain obligations on the parties with regard to consolidating their efforts and the means necessary for implementing specific water protection measures with an indication of the terms of their fulfillment. The drafting of basin agreements is carried out on the basis of their water management balances, schemes of integrated water bodies/facilities use and protection, governmental economy programmes, scientific and design works, and forecasts of socioeconomic development. In order to achieve the objectives and goals stipulated by basin agreements, physical and legal entities may establish foundations for the purpose of implementing measures on water bodies/facilities rehabilitation and protection. Corresponding river basin water management organizations are tasked with the preparation and implementation of basin agreements.

Clause 3 of Article 63 of the Water Code of the Republic of Kazakhstan stipulates that public associations, on their own initiative, may perform the functions of social control in the field of water fund use and protection. To this end, the order of conducting social control is determined by public associations, in coordination with state bodies charged with a mandate of state control in the field of water fund use and protection.

One of the forms of enhancing public participation in water resources management is the establishment and expansion of water user associations' networks. For this purpose, Kazakhstan adopted the Law "*On rural consumers' cooperatives of water users*" in April 2003, according to which a cooperative is a voluntary association established by individual citizens and/or legal entities for the purpose of joint operation and joint management of water use systems in the interests of agricultural needs. A cooperative of water users carries out the planning of water delivery and distribution schedules within the limits of set water-use quotas, performs the operation and reconstruction of the water use system and elements, and conducts construction, repair, cleaning and works. In essence, water-user associations substantively participate in the process of water resources management.

Thus, at present, positive tendencies have taken shape in terms of more active public participation in the process of water resources use and protection. In addition, a legal basis has been developed and conditions are being improved, which facilitate the involvement of the population and of legal entities in addressing the vital problems of the region.

WATER PRODUCTIVITY INCREASE—THE MAIN GOAL OF IWRM AND WAYS TO OVERCOME POVERTY

Sh. Muhamedzhanov
Scientific-Information Centre ICWC, Tashkent, Uzbekistan

The Central-Asian region, including five former Soviet republics, is located in an arid zone with severe freshwater scarcity. Settlements and irrigated lands along the major rivers (Amu-Darya and Syr-Darya), in both mid-stream and downstream sectors, face acute water deficits, even in humid years. The region's economy is mostly agrarian. Nearly 70% of the population live in rural areas and their well being is determined by agricultural production. In turn, agriculture depends on water resources and their effective utilization.

A common irrigation system has developed across Central Asia. Large irrigation systems transcend the irrigated massifs of different states. Water resources use and management falls under the jurisdiction of five states: Kazakhstan, Kyrgyzstan, Tajikistan, Turkmenistan and Uzbekistan. Each state faces difficulties, both in terms of their internal systems and in terms of interstate issues, since each country is at a different stage of water and agricultural sector reform.

Central-Asian regional development, with regard to each individual state's interests, becomes impossible without a single strategy and approach to water sector reform aimed at improving water and irrigation management. Presently, much water is used for irrigation. In some regions, under conditions of low main crop yield, productivity does not cover the expenses incurred in delivering the water to that region. The main criterion of irrigation water productivity should be maximum yield under minimum water expenses. The main advantage of water sector reform is that it creates responsibility for water, which will limit excessive water use.

To achieve rational water use, irrigation management and effective water allocation, reforms in the irrigation sub-sector must be built on a participatory principle, which should include water users and water suppliers. Modes of reform have been adapted on pilot sites, with dissemination of these results leading to further recommendations.

SIC ICWC, working jointly with IWMI within the project "IWRM-Ferghana," has considered issues of irrigation water and irrigated land productivity increase and the improvement of water resources management. Much of the work has centred upon establishing a structure of effective water

P. Wouters et al. (eds.), Implementing Integrated Water Resources Management in Central Asia, 95–103.

allocation and of optimal irrigated farming performance with minimum irrigation water losses. To tackle these issues, irrigation water use monitoring and actual irrigation water and irrigated land productivity evaluation and recommendations development have been organised since 2002. With this purpose in mind, 10 demonstration plots were selected in the Gulyakandoz Canal in Sogd oblast of Tajikistan, the South-Ferghana Canal in Ferghana and Andizhan oblast of Uzbekistan, and the Aravan-Akbura Canal in Osh oblast of Kyrgyzstan. Demonstration plots were selected in the head, middle, and tail part of canals. Each demonstration plot within the selected canals was chosen taking into account its representative value for the whole canal command area. These plots also cover the different climatic zones of the Ferghana valley.

The crop pattern in the Ferghana varies in different regions of the valley. In the Uzbek and Tajik sections, cotton covers 40%, and wheat 30%. In Osh oblast, cotton covers only 7%, and wheat 33%. Crop patterns of farms tend to fit into the oblast pattern, which dictates the dominant crops.

Assessment of Actual Irrigation Water Use on Demonstration Plots

Analysis of actual irrigation in 2002 showed that, with variations due to soil-reclamation conditions, relief, and land levelling, there are significant irrigation water losses in farms. In some farms, losses are due to the selection of the wrong irrigation scheme and duration.

In Sogd oblast, on all three demonstration plots, the durable irrigation period was found to run up until October. Irrigation of small plots with short furrows is typical of the region. A difference in irrigation water use regarding main canal location has been noticed. Farms in the head part of the canal, with higher water availability, use more water compared to farms in the middle and tail sections of the canal. Big losses in water release are found in farms located in the head part of the canal.

In the Ferghana and Andizhan oblasts, irrigation conditions differ between plots depending on soil-reclamation and hydrogeological conditions. In farms located on highly permeable soils, with unstable water availability and poorly selected irrigation schemes, high water losses were observed. The largest part of supplied water (41%) is lost through percolation. Besides natural factors, irrational water and land use leads to high water losses. In Nozim farm in the Ferghana oblast, bad field levelling and durable irrigations of up to 15 days, allied to durable irrigation intervals of up to 30 days in long furrows (5252 m), led to high water losses.

Where water release losses were absent, the main losses occurred due to infiltration. In Turdali farm, in the Ferghana oblast, irrigation norms and

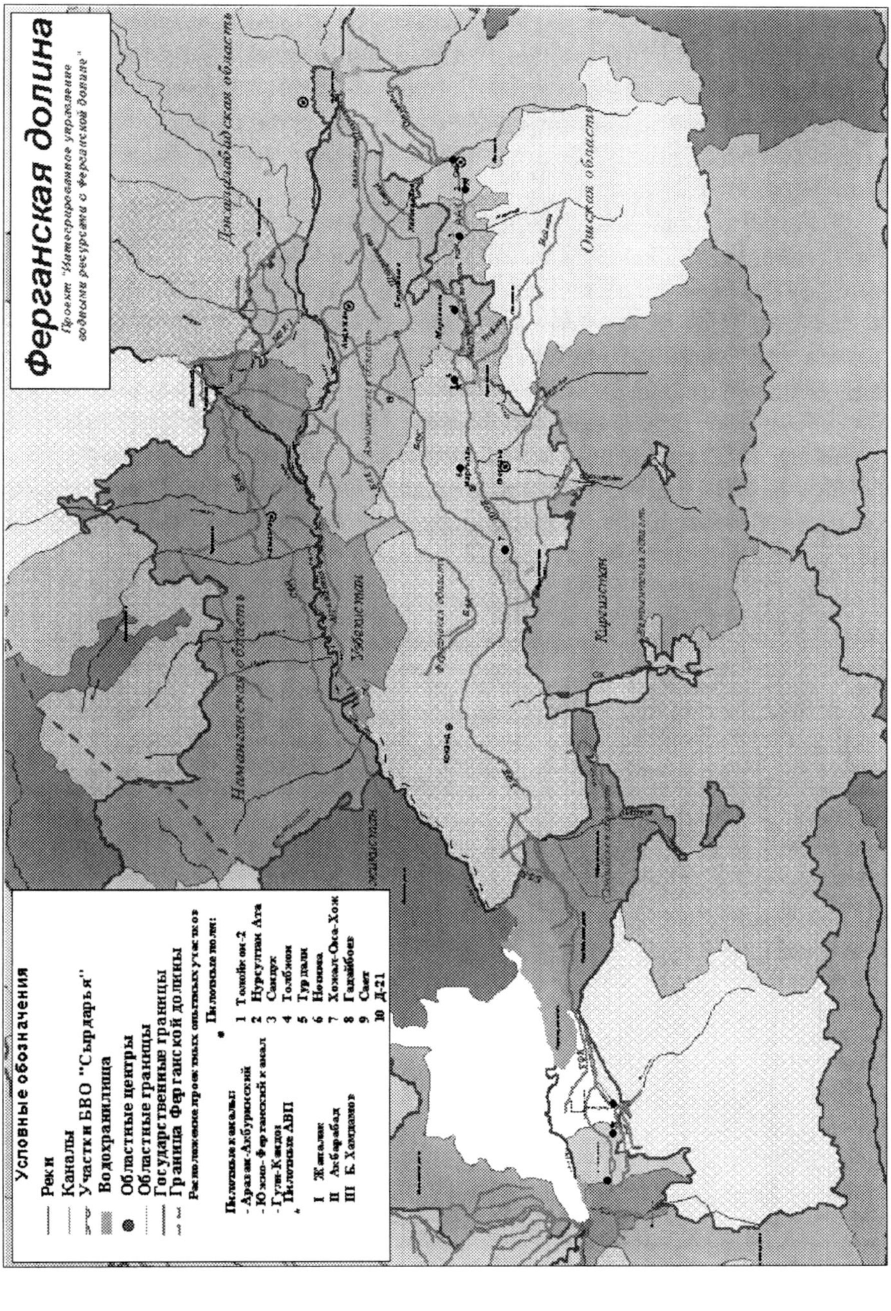
Ферганская долина
Проект "Интегрированное управление водными ресурсами в Ферганской долине"
Условные обозначения
Реки
Каналы
Участки БВО "Сырдарья"
Водохранилища
Областные центры
Областные границы
Государственные границы
Граница Ферганской долины
Наманганская область
Узбекистан
Кыргызстан
Ошская область

discharges (0.3–0.4 l/s) were optimal for farm conditions (ground water depth was less than 0.5 m). In Tolibjon farm, in the Andizhan oblast, irrigation was performed only on local plots where crops needed water. In Osh oblast in Kyrgyzstan, irrigation management was aggravated in all three farms by complicated relief and stony soils. Farmers in the Toloikon and Nursultan-Ali farms achieved a desirable level of soil moistening by applying the high irrigation norm. The largest part of supplied water was spent in water release from irrigated fields.

The analysis of results from the test farms show that the largest losses occur due to percolation. Actual losses in farms exceed the norm. Indeed, in some farms, losses from percolation are double the norm. In these farms, efficiency is only 40%. High losses from percolation are inevitable for most farms with high soil permeability, long furrows and bad levelling. However, it is worth noting that separate farms with complicated soil-reclamation conditions have low losses from percolation and release. These farms have high irrigation efficiency. This is achieved due to short furrows and small discharge to these furrows, as well as effective use of ground water contribution. However, with these notable exceptions, irrigation efficiency on the demonstration plots is very low, as demonstrated in Table 1, below:

TABLE 1. Main indicators of irrigation water use on demonstration fields

Farm	N m^3/ha actual	Actual losses for release, %		Actual losses for percolation, %		Field efficiency, % actual	Ea
		norm	actual	norm	actual		
Tajikistan							
Brigade № 21	8,266	12.5	10.3	11.7	19.7	70	0.70
Sayed	7,343	16.9	20.8	20.2	20.2	59	0.59
Gadoiboev	12,969	16.9	19.5	20.2	35.5	45	0.45
Uzbekistan							
Hodzhahon-ona -Hodzhi	16,795	12.9	18.2	17.4	40.6	41	0.41
Nozima	6,718	1.9	0	30.9	58.1	42	0.42
Turdiali	2,145	10.3	5.1	11.5	10.7	84	0.84
Tolibjon	9,510	16.9	12.9	20.2	28.5	58	0.59
Kyrgyzstan							
Toloikon	5,803	1,3	32	45,8	40.2	28	0.28
Nursultan-Ali	5,120	4,9	18,4	26,9	31.2	50	0.50
Sandik	6,030	1,3	25,8	45,8	10.7	64	0.64

where: Ea = (N – Ссбр – Winf)/N; N – water supply to field, m^3/ha; Ссбр—release from irrigated field, m^3/ha;
Winf—потери на глубинную, фильтрацию m^3/га.

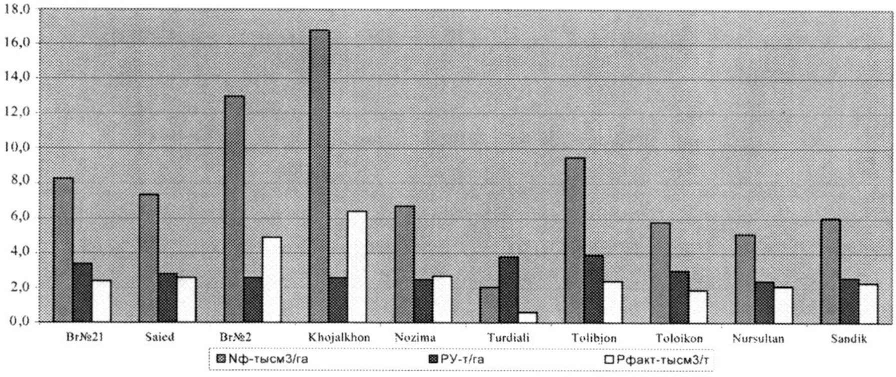

Figure 1. Assessment of actual irrigation water productivity

Assessment of Actual Water Productivity on Demonstration Plots

To evaluate irrigation water productivity, and to analyse and assess irrigation water use and agro-technical measures, field-monitoring results were taken on the demonstration plots for the growing season. Comparative assessment of irrigation water use shows that the actual water volume supplied to the fields is less than the required volumes, and that water productivity increase is possible only due to irrigation volume and number reduction.

In Tajikistan (Sogd oblast), actual irrigation water productivity amounted to 2.4–4.4 th m^3/t, as illustrated in Figure 1, below:

If we take into account the fact that irrigation performed in September and October does not have any effect other than to reduce crop yield, we can see that water productivity would be much higher without these irrigations, and would amount to 1.9–2.6 th. m^3/t, as illustrated in Table 2, below. The highest expenses per production unit are found in farms with highly permeable soils; there, productivity is reduced by as much as 2.5 times. The lowest irrigation water expenses, and correspondingly the highest productivity, were achieved in those farms where irrigation water was used rationally and where agro-technical measures were performed in a timely fashion.

Besides unproductive water losses through percolation and release from the field, water productivity reduction is connected with yield losses. On the basis of field monitoring results, the different factors contributing to yield reduction can be determined. Most yield losses occur due to low humus content in the soil. In farms in the Osh oblast, losses due to lack of humus amounted to between 30% and 40%. In farms in the Sogd oblast, these losses ranged from 11% to 23%. In farms in the Ferghana and Andizhan oblast, relatively high levels of humus content explain losses of less than 10%. Soil salinization is one of the most important factors. On saline lands, yield loses amount to 9–13%,

TABLE 2. Main indicators of irrigation water use on demonstration fields

Farm	N gross Actual m³/га	Nopt gross	Losses for release Norm %	Losses for release Actual %	Losses for percolation Norm %	Losses for percolation Actual %	Py	Пу	Ex opt	Exp
Brigade № 21	8,266	6,642	12.5	10.3	11.7	19.7	3,2	4.9	1.9	1.4
Sayed	7,343	7,296	16.9	20.8	20.2	20.2	2.8	5.6	2.6	1.3
Gadoiboev	12,969	7,587	16.9	19.5	20.2	35.5	2.4	4.9	2.9	1.5
Hodzhahon-ona-Hodzhi	16,795	8,038	12.9	18.2	17.4	40.6	2.6	6.1	3.1	1.29
Nozima	6,718	4,074	1.9	0	30.9	58.1	2.4	6.5	1.6	0.6
Turdiali	2,145	2,090	10.3	5.1	11.5	10.7	3.5	5.4	0.55	0.4
Tolibjon	9,510	6,871	16.9	12.9	20.2	28.5	3.8	6.3	1.8	1.1
Toloikon	5,803	2,982	1.3	32	45.8	40.2	3.0	7.6	1.0	0.4
Nursultan-Ali	5,120	3,530	4.9	18.4	26.9	31.2	2.4	7.2	1.5	0.5
Sandik	6,030	7,072	1.3	25.8	45.8	10.7	2.9	5.4	2.7	1.3

Note: Nopt—optimal irrigation norm; Nфакт—actual irrigation norm; Py—actual crop yield; Пу—potential crop yield; Ex opt—optimal water expenses per production unit; Exp—potential water expenses per yield unit.

with a maximum of 10–18.7%. Other factors influence yield to a lesser extent. Without taking into account yield losses due to agro-reclamation factors, any assessment of potential water productivity would be inaccurate, since under optimal irrigation duty, low yield due to factors not relating to water will lead to low productivity value.

The optimal gross irrigation norm and potential yield for each demonstration field were used to calculate potential water productivity (see Table 3, below). Productivity in the demonstration fields, under the elimination of all reducing factors, could be increased by 54% in Tajikistan, 52% in Uzbekistan, and 34% in Kyrgyzstan.

Economic Indicators of Irrigation Water Use Productivity

The main indicator of agricultural production efficiency is income from grown production, defined by total expenses for production and the amount of production and its realisation. Total expenses for production and the amount of production depend on different factors, including irrigation water use.

As a result of the monitoring project, farm's expenses for production were defined and analysed for each demonstration field. Total expenses in local currency were calculated, taking into account work volume and costs. To allow

TABLE 3. Main indicators of irrigation water use productivity

Farm	Nф th. m³/га	Nopt th. m³/га	Py th/га	Пу th/га	Exф th. m³/т	Exopt th. m³/т	Exp th. m³/т	Pr-fact th/th. m³	Pr-opt th/th. m³	Pr-poten th/th. m³
Brigade № 21	8.27	6.64	3.2	4.9	2.4	1.9	1.4	0.4	0.51	0.74
Sayed	7.34	7.30	2.8	5.6	2.6	2.6	1.3	0.4	0.38	0.77
Gadoiboev	12.97	7.59	2.5	4.9	4.4	2.9	1.5	0.2	0.34	0.65
Hodzhahon-ona-Hodzhi	16.79	8.04	2.6	6.1	6.4	3.1	1.3	0.2	0.32	0.76
Nozima	6.72	4.07	2.4	6.5	2.7	1.6	0.6	0.4	0.61	1.59
Turdiali	2.145	2.09	3.5	5.4	0.6	0.55	0.4	1.8	1.8	2.58
Tolibjon	9.51	6.87	3.8	6.3	2.4	1.8	1.1	0.4	0.57	0.92
Toloikon	5.80	2.98	3.0	7.6	3.9	1.0	0.4	0.5	1.0	2.55
Nursultan-Ali	5.12	3.53	2.4	7.2	2.1	1.5	0.5	0.5	0.68	2.04
Sandik	6.03	7.07	2.9	5.4	2.3	2.7	1.3	0.5	0.37	0.76

Nф; Nopt—actual and optimal irrigation norm.
Py; Пу—actual and potential yield capacity.
Exф, Exopt, Exp—actual, optimal and potential irrigation water expenses.
Pr-fact, Pr-opt, Pr-poten—actual, optimal and potential water productivity.

TABLE 4. Irrigation water and production cost

Farm	Ewf $/ha	Ewcr $/t	Cost of 1t production th. н.в.	Cost of 1t production $
Brigade № 21	12.2	3.8	0.5	149
Sayed	11.2	4.1	0.5	163.3
Gadoiboev	14.3	5.8	0.5	180.3
Hodzhahon-ona-Hodzhi	0	0	145	140.8
Nozima	0	0	145	140.8
Turdiali	0	0	145	140.8
Tolibjon	0	0	144.2	140
Toloikon	2.9	1	5.1	111.6
Nursultan-Ali	3.5	1.4	5	109.4
Sandik	3.4	1.2	6.9	151.3

Ewf—total cost of water, Ewcr—cost of water spent per 1t production.

for a comparative assessment, total expenses were converted into US Dollars. The largest portion of expenses was attributed to manual labour, machinery and fertilizers. Farms owning their own machinery spent less money for their operation than farms leasing machinery. Nursultan-Ali farm in Kyrgyzstan incurred no expenses for manual labour because family members perform all work on the farm.

Expenses for water constitute 5% of the total expenses, but in some farms they are higher than expenses for seeds, transport and agro-chemicals. For Uzbekistan, expenses for water are not indicated because it is provided free

TABLE 5. Assessment of agricultural production and used irrigation water productivity

Farm	Nф m^3/ha	Realization 1,000 $	Expenses 1,000 $	Expenses $ha	Profit $tт	Profit $/t	Water price Local currency /m^3	Water price $/$m^3$
Brigade № 21	8,266	2.9	1,7	280	62	199	0,2	0.06
Sayed	7,343	1.8	1.5	386	25	69	0.2	0.06
Gadoiboev	12,969	5.6	4.1	317	46	114	0.1	0.03
Hodzhahon-ona-Hodzhi	16,795	1.9	1.4	272	38	99	22.8	0.02
Nozima	6,718	2.7	1.6	199	59	142	52.3	0.05
Turdiali	2,145	1.3	0.8	268	56	197	248.7	0.24
Tolibjon	9,510	2.7	1.7	336	51	195	57.5	0.06
Toloikon	5,803	1.3	0.9	214	41	121	2.6	0.06
Nursultan-Ali	5,120	0.2	0.1	159	44	109	2.4	0.05
Sandik	6,030	2.2	1.3	268	57	165	3.3	0.07

of charge. The total cost of agricultural production amounts to 159–386 \$/ha, with the cost of 1t production being 64–138 \$/t. Total income from production of cotton on demonstration fields amounted to 140–180 \$/t. For winter wheat the figure was 112 \$/t and for summer wheat, 109 \$/t.

The total cost of water per 1 t of production was calculated. For farms in Tajikistan and Kyrgyzstan, the figure is 1–5.8 $\$/m^3$ (see Table 4, below).

The net profit from production on demonstration fields amounted to 25–62 \$/t, as shown in Table 5, below. On the basis of production and spent irrigation water amounts, irrigation water productivity analysis and economic assessment was calculated. Its value amounted to 0.02–0.24 $\$/m^3$.

Field monitoring showed that all the farms in the Ferghana valley had low water and land use efficiency. In the case of input provision, free prices on agricultural production, and control over water use, there are preconditions for agrarian sector development at state and interstate level, which make possible high crop yield and an increase in irrigation water and land productivity.

IMPLEMENTATION OF THE ARAL SEA BASIN SOCIOECONOMIC MODEL: AN ASSESSMENT OF THE OPPORTUNITIES TO BE GAINED THROUGH REGIONAL ECONOMIC INTEGRATION

Makhmud T. Ruziev and Valeriy G. Prikhodko
Advanced Specialists of the Scientific Information Centre of the Interstate Commission for Water Coordination (SIC ICWC) of Central Asia.
E-mail: mv@icwc-aral.uz

Forecasting future development in any region is a very complicated process, even in developed societies. Although past trends can be used as the basis for predicting the future, combinations of trends can cause qualitative changes in predicted trends and trend dynamics (see, for example, D. Forester). Forecasting under transitional economic conditions (as is the case in the Aral Sea Basin) is more complicated as the political systems are changing and environmental degradation is increasing. It is, therefore, almost impossible to forecast the definite future conditions of the five Central Asian States which used to be under the tough political control of the USSR, or of their unpredictable neighbor, Afghanistan.

Many ways have been suggested to assess progress toward sustainability based on the implementation of computer models. This paper explains the concept of the Socioeconomic Model and examines the results of its implementation in terms of forecasting different future development scenarios.

1. Introduction

Assessment, of the future from the point of view of sustainability has always been important in terms of defining a long-term development strategy. Such assessment can be made through mathematical models simulating demographic, economic, ecological, and other processes. They do not answer directly whether we will live in better or worse conditions in the future but they may give indicators or suggest criteria to assess the economic, social, and ecological well being of the state. Nowadays, computers are powerful enough to calculate and show the results of planned human actions. The software to support such models has to be specially developed. The Socioeconomic Model (SEM) is one such example.

P. Wouters et al. (eds.), Implementing Integrated Water Resources Management in Central Asia, 105–123.

The development of the SEM was a product of the adaptation of the "Globesight" model to the specific conditions and parameters of the Aral Sea basin. The "Globesight" model was created in Case Western Reserve University under the supervision of Professor M. Mesarovich and was presented to the SIC ICWC workgroup at 1999, within the framework of the UNESCO initiative "Aral Sea Basin Vision."

Further development of the SEM resulted from the creation of the "Aral Sea Basin Management Model" (ASB-MM). ASB-MM was created through the WEMP program for the "Central Asian Interstate Sustainable Development Commission" by "Resource Analysis" (in Delft, The Netherlands) and SIC ICWC (in Tashkent and Uzbekistan). These parties worked under the aegis of the "UNDP Aral Sea Basin Capacity Development Project" and the "International Fund for Aral Sea Saving."

2. Socioeconomic Model

The philosophy of the model is "Vision," which is neither forecast nor prediction but rather defines a desirable, achievable future. "Vision" can contribute to the changing of tendencies, can strengthen them (if they promote the achievement of "Vision"), and can respond to them in order to provide the desired future outcomes. "Vision" can create new tendencies and prevent the occurrence of undesirable ones. Planning on the basis of "Vision" calls for backward analysis before starting strategic development. In other words, it is first of all necessary to define "where we want to go" and then to correlate this with "where we are."

While defining "where we want to go," we must decide what changes are required to current relations, practices, and tendencies in order to achieve the "Vision." These changes, in turn, are made through the targeting of specific interim goals. The goals, for their part, form the basis for the strategies put in place in order to tackle the problem of "how to reach the set aims."

Schematically, the process can be represented as:

Vision → Changes → Objectives → Strategy → Plans → Actions

SEM attempts to predict what will happen within the socioeconomic system. Human behavior is very important here. Although this factor is not very predictable, it is possible to extrapolate some current figures and apply them to future situations. The model takes such an approach in terms of (see Figure 1):

- population (rural and urban);
- economy (GNP, GNP per capita, contribution of economic sectors to GNP);
- water (demand of economic sectors; availability is calculated by hydrological model);

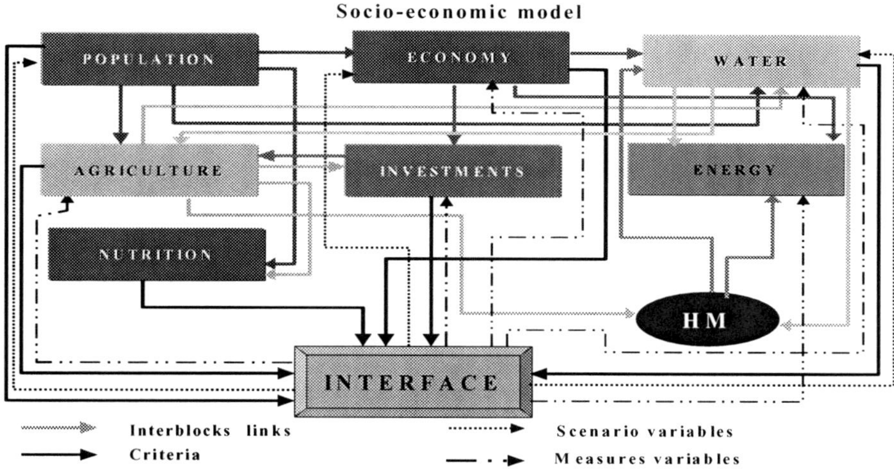

Figure 1. Socio-Economic model structure.

- agriculture (irrigated land productivity by crops);
- investments (investments in agriculture, direct foreign investments, investments in land reclamation, investments in new land development);
- power (production and consumption);
- nutrition (calorie production and consumption, with regard to food basket);
- impact of climate change on water resources and agricultural production, etc.

An examination of the construction of SEM and the general principles of its functioning is given in below. The work was divided into several stages, differing from each other in terms of complexity and the number of parameters involved. The latest version of SEM works in the following way:

1. The model sets a series of STATE parameters $A_i(2000)$ ($i = 1$–40), such as Population in 2000, GNP in 2000, Irrigated lands in 2000, Specific cost of new land development, etc. $A(2000)$ parameters are fixed and can be changed only by the model's authors.
2. The user can apply a set of MEASURES $M^j_K = 1$–15, such as industry or agriculture share in GDP, investment climate, water saving, hydropower share, land improvement, etc. and can choose one of three SCENARIOS $S^j_P = 1, 2, 3$ (with different GDP growth, population growth, and climate changes impact).
3. As a result of the selected MEASURES application and SCENARIO choice, and after the running of SEM and HM, the SEM system turns from the initial STATE $A_i(2000)$ to the final STATE A^j_i (2020) (Figure 2). The set of MEASURES $M^j_K = 1 - 11$, SCENARIOS $S^j_P = 1, 2, 3$, and final STATE parameters A^j_i (2020) form the CASE_j

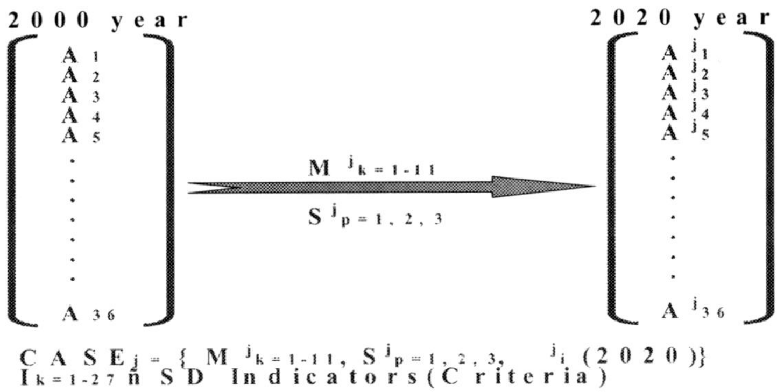

Figure 2. General principle of functioning of SEM.

4. The INTERFACE of the model allows the user to save different CASES$_j$ and compare them with each other by giving priority to one or other of the sectors (socioeconomic development, water use, agriculture, environment) (see Figure 3). In order to identify the most sustainable solutions or measures, the user could use SD Indicators $I_K = 1–27$ (such as GNP per capita or, calories export/import) (see Figure 4).

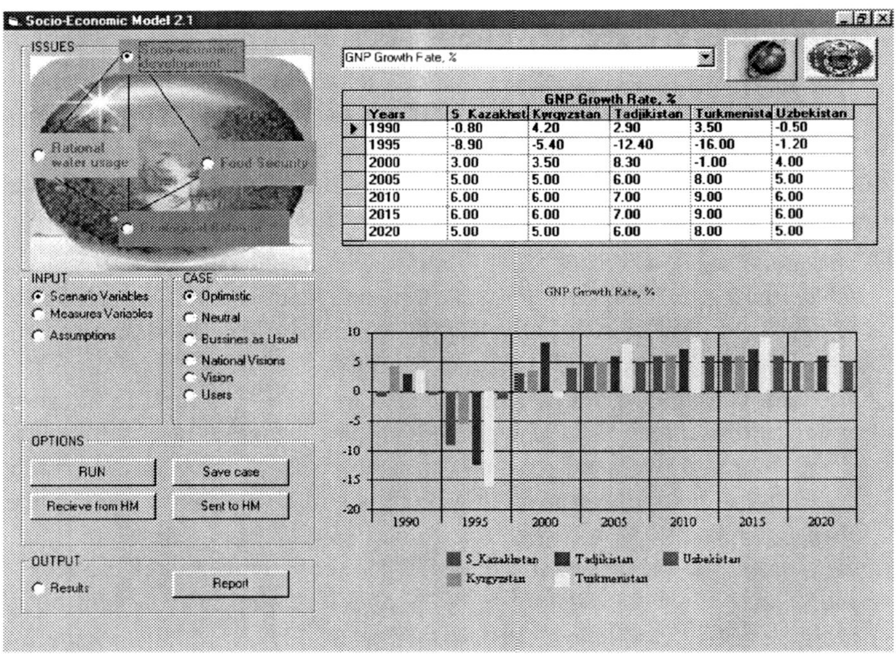

Figure 3. Interface of SEM.

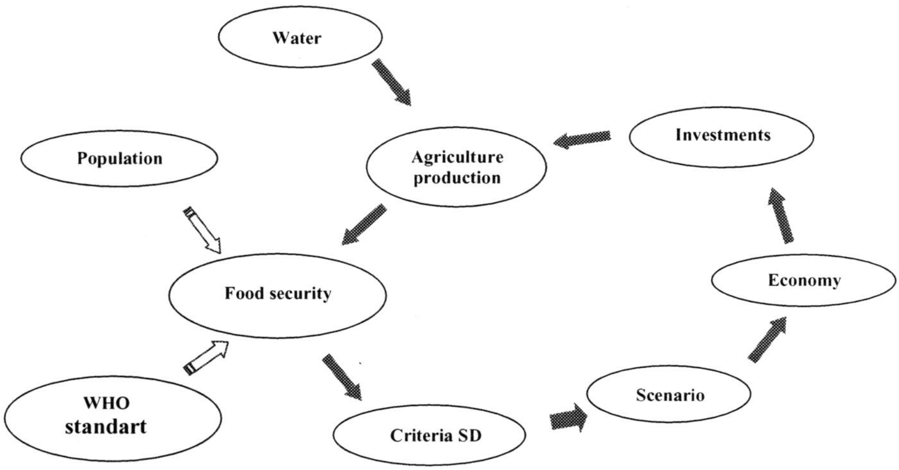

Figure 4. SEM main casual loops.

Five possible future Aral Sea basin region scenarios were developed to explore some of the above-mentioned issues and to assess their relative importance in the different states of the region (see Figures 5, 6, and 7).

1. *Optimistic scenario*
   - The region will be developed through the improvement of integration processes that are today being developed and planned by the governments of all countries. Such developments include:
     - mutually beneficial sharing of all transboundary water resources, based on water saving and common environmental approaches;
     - mutually beneficial development of the agrarian sector, with the focus on sharing regional production, particularly in respect of the most favorable cropping patterns;
     - coordinated processing in the agrarian sector and its infrastructure.
   - Population growth rates are will decrease, by 2020, to 0.98% per year, thus meaning that the population will amount to 54 million people. Average annual GNP growth will be 4–6% in the period 2000–2010, around 6% from 2010 to 2015 and less than 5% from 2015 to 2020. GNP in the region is expected to be about 86 billion US$, that is more than 1600 US$ per capita/year.
   - Power engineering will be developed mainly on the basis of hydroelectric power plants, so as to create a steady supply of environmentally friendly power generation.
   - It is predicted that water saving policy implemented at national level will allow achievement of the following parameters of water use efficiency:

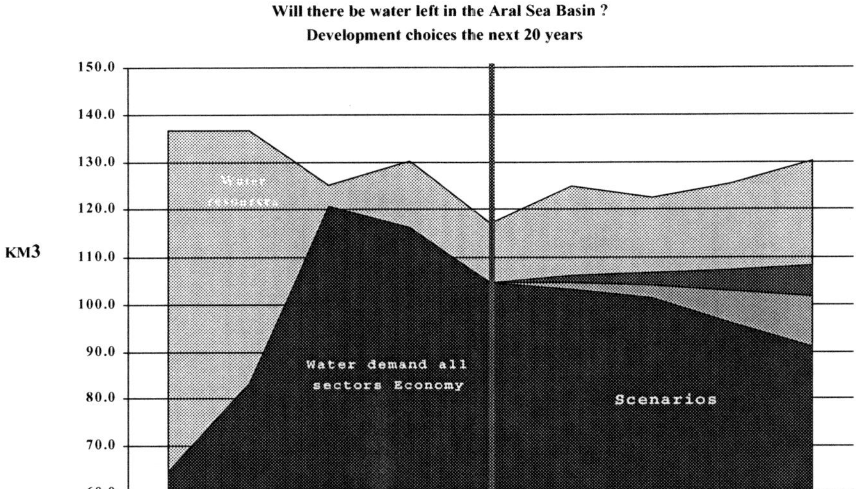

	1960	1970	1980	1990	2000	2005	2010	2015	2020
Water resources	136.6	136.7	125.0	130.3	117.0	124.8	122.3	125.5	130.2
Business as Usual	64.7	83.5	120.7	116.1	104.6	106.2	106.7	107.4	108.4
Neutral	64.7	83.5	120.7	116.1	104.6	104.6	104.2	103.0	101.8
Optimistic	64.7	83.5	120.7	116.1	104.6	103.2	101.4	96.1	91.1

Figure 5. Past trends and prediction (by different scenarios of development): Water Demand.

unit water consumption in irrigation—9.4 thousand m^3/ha; domestic water consumption—0.08 m^3/man/year.

- A set of measures aimed at the decrease of water consumption in irrigation will lead to an increase of irrigated lands to 8.5 billion ha, against the present level of 7.85 billion ha. This increase of irrigated lands is expected primarily after 2010, in that it is connected to the expected general improvement of the economic situation in the region and the accumulation of sufficiently large funds to implement large-scale water saving measures. Furthermore, implementation of measures for the increase of agricultural production efficiency up to 80% will improve food supply to the population. Average production is expected to reach 3500 Kcal /person/day under prevailing vegetable and fruit rations. An optimal combination of food production and cash crop production will allow reductions in the import of grain and dairy products and the increased export of vegetables, fruits and other products.
- A significant increase in GNP will occur through rapid industrial growth. Taking into account industrial development on the one hand, and the application of water recycling in industrial plants on the other, water use in industry is forecast to achieve 3.3 billion m^3/year, against 1.9 billion m^3 in 2000.

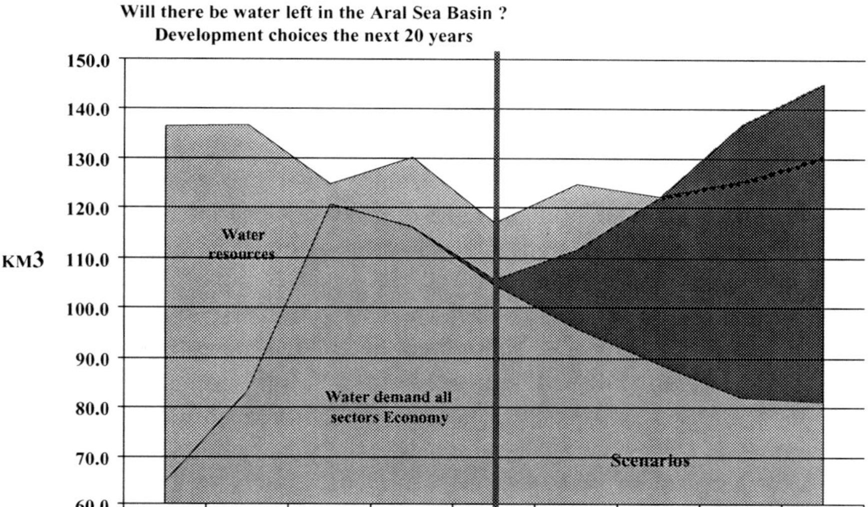

	1960	1970	1980	1990	2000	2005	2010	2015	2020
□ Water resources	136.6	136.7	125.0	130.3	117.0	124.8	122.3	125.5	130.2
□ Nations vision	64.7	83.5	120.7	116.1	105.6	111.5	122.1	136.8	145.3
□ Vision UNESCO	64.7	83.5	120.7	116.1	104.6	95.8	88.9	81.9	81.3

Figure 6. Past trends and prediction (by different scenarios of development): Water Demand.

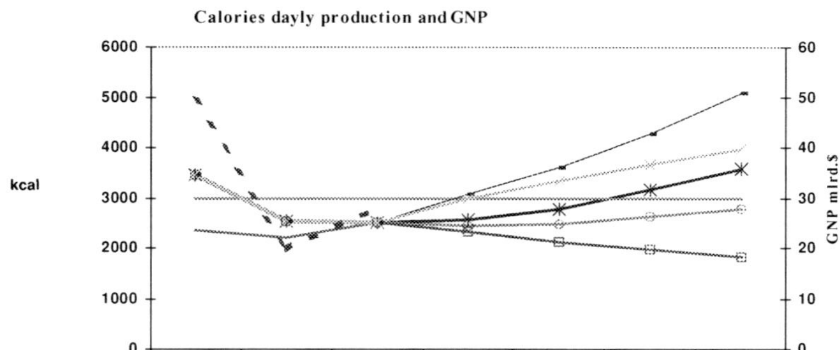

	1990	1995	2000	2005	2010	2015	2020
Real calories consumption	2345	2199	2510				
Optimistic	3449	2538	2500	2569	2789	3164	3587
Neutral	3449	2538	2500	2454	2490	2624	2773
Business as Usual	3449	2538	2500	2316	2134	1967	1818
1990-2000	3449	2538	2500				
WHO Standart	3000	3000	3000	3000	3000	3000	3000
National Vision	3449	2538	2500	3064	3600	4271	5081
Vision UNESCO	3449	2538	2500	2990	3345	3658	3971
GNP	49.4	19.7	27.9				

Figure 7. Past trends and prediction (by different scenarios of development): Calories daily production and GNP.

- Given the potential water use efficiency in different economic sectors, total water consumption will be 91.1 km^3/year. At the same time, 80.1 km^3/year will be used in irrigation and 11 km^3/year will be used in industry and the household sector.

2. *Neutral scenario*
   - Integration processes in transboundary water management will be developed more slowly than in the optimistic scenario. Additionally, this scenario does not assume regional crop specialization and coordinated processing in agriculture.
   - Population growth rates will decrease slightly, reaching 1.44%/year by 2010 and 1.23%/year by 2020. Thus, by 2020, the population will amount to 55 million. GNP will increase by 2–4 % per year. In 2020, the GNP for the region is expected to be around 62 billion US$, or 1133 US$ per capita.
   - Development of new lands is restricted not only by water availability and quality, but also by lack of necessary investment. Taking into account the fact that the given scenario proposes low economic development and limited financial resources for implementing water saving measures in all economic sectors, water use efficiency parameters will be as follows: unit water consumption in irrigation—11 thousand m^3/ha; domestic water consumption—0.09 m^3/person/year. Irrigated lands will increase by approximately 500,000 ha, reaching 8.4 million ha by 2020.
   - Given the planned water use efficiency in this scenario, total water consumption will be 101.8 km^3/year. Irrigation water requirements are expected to be 91 km^3/year, with totals of 2.5 km^3/year and 4.9 km^3/year for the industrial and household sectors respectively.
3. *Business as usual scenario*
   - The region will be developed along the lines of current tendencies in transboundary water use, and in the field of regional agricultural integration, both in agricultural production and processing. The main efforts of the states will be directed toward saving local water sources.
   - Population growth rates will remain constant at a level of 1.9% per year, meaning that the population will be around 61 million by 2020. The average annual increase in GNP will not exceed 4% per year. At the same time, regional GNP is expected to be 44.7 billion US$, that is around 800 US$/capita/year.
   - According to current tendencies, water use efficiency parameters are expected to be as follows: irrigation—12 thousand m^3/ha; domestic—0.1 m^3/person/year.
   - By 2020, the total irrigated area will not change from current levels.

- Total water consumption will be 108.4 km^3/year. Thus, 96 km^3/year will be used in irrigation, while the industrial and household sectors will use 3.05 km^3/year and 6 km^3/year respectively.

4. *Scenario based on national strategies/plans of long-term development of the Aral Sea basin states.*
   The basic parameters of national strategies, visions and long-term development plans used in the model calculations are shown in Table 1, below.
5. *Scenario based on "Water Vision of Aral Sea Basin for 2025," Department of Water Sciences, UNESCO, 2000.*

3. Results

3.1. OPTIMISTIC, NEUTRAL AND BUSINESS AS USUAL SCENARIOS

In the first scenario, potential daily production of calories per capita will exceed 3000 calories per day by 2010. Thus, there will be an opportunity in the region to produce more food than is required. In the second scenario, food production would reach only 2700 calories per person per day. In the third scenario, a sharp decrease in food production is forecast, particularly after 2005. This is expected to be the situation in the region as a whole. However, in some countries of the Aral Sea basin, the situation will be slightly different from the overall predictions. So, according to the calculations over the whole forecast period, Tajikistan will not be able to ensure food production of 3000 calories per person per day, whilst Uzbekistan and Kyrgyzstan will reach this level by 2014 and 2015 respectively. However, for the region as a whole, food production balance will be positive after 2010. Thus, food production in the region as a whole will be sufficient to meet the population's needs. So, for example, by overproducing meat in Kazakhstan, Kyrgyzstan, and Turkmenistan, it will be possible to meet the requirements of the entire region for this product. It should be taken into account that we consider production volume only for the Aral Sea. The same situation relates to cereal production. Thus, the model calculations prove once again the need for the integration of agricultural production in the whole region.

Under present conditions, the full potential of irrigation development will not be realized by the end of the forecast period under any of the scenarios. However, it can be concluded that food demands may be met without the large-scale development of irrigation. In order to have such an opportunity in the second and third scenarios, agricultural production efficiency must be increased, and this is one of the possible directions for decision-makers to address.

TABLE 1. Basic parameters of national strategies, visions, and long-term development plans

Parameters	Years	Edit	S. Kazakhstan	Kyrgyzstan	Tajikistan	Turkmenistan	Uzbekistan
Population growth rate	2010	%	1.30	1.50	4.20	3.20	1.95
	2020	%	1.30	1.50	3.00	3.20	1.95
Population	2010	Million people	2.77	2.72	8.84	8.52	30.08
	2020	Million people	3.16	3.15	12.15	11.68	36.48
GNP	2010	Billion $	6.5	2.4	2.9	12.2	33.8
	2020	Billion $	11.3	4.2	5.5	64.1	65.2
GNP growth rate	2010	%	6.0	6.0	6.1	18.0	8.0
	2020	%	5.0	5.0	6.0	18.0	6.5
Agricultural Contribution to GNP	2010	%	34	50	30	15	24
	2020	%	32	50	30	15	24
Contribution industry in GNP	2010	%	24	20	26	32	25
	2020	%	26	22	26	32	25
Irrigated lands	2010	1000 ha	809.5	447.5	822.6	1897.5	4712.9
	2020	1000 ha	881.8	479.7	959.7	2343.9	4915.0
Irrigated lands per capita	2010	ha/person	0.29	0.16	0.09	0.22	0.16
	2020	ha/person	0.28	0.15	0.08	0.20	0.13
Domestic specific consumption	2010	1000 m^3/person/year	0.06	0.04	0.08	0.07	0.09
	2020	1000 m^3/person/year	0.07	0.06	0.08	0.08	0.09

The most sustainable parameter is irrigated area per capita. Although it is being reduced, this is natural since the rate of population growth exceeds that of the irrigated area, and as such this parameter remains within the marked limits.

Daily calorie consumption, according to the first scenario, is more than 3000 calories per day in all the states, excluding Tajikistan. According to the second scenario the opportunity to produce food (less than 3000 calories/person/day but more than 2000 calories/person/day) is reduced in the Kyrgyz Republic and in Uzbekistan. According to the third scenario, Turkmenistan will also have such reduced opportunities. For the region as a whole, under the first scenario it is possible to completely meet daily food consumption at more than 3000 calories, whilst according to the second and third scenarios this daily consumption varies between 2000 calories/person/day and 3000 calories/person/day.

According to the optimistic scenario, the amount of used water (along with the Aral Sea's water requirements) does not exceed 90% of available water. At the same time, the population may use 220 l/day/person, while head water intake for irrigation will be 9.4 thousand m^3.

In the neutral scenario, water intake is slightly increased due to population growth. Water use is around 95% of available resources, while specific water intake for irrigation is around 10.7 thousand m^3.

3.2. NATIONAL PLANS/STRATEGIES OF LONG-TERM DEVELOPMENT

According to national plans of long-term development, by 2020 the population in the region will grow to 65 million. Considerable increases in population numbers are expected in Tajikistan and Turkmenistan where populations are expected to be 8.6 and 11.68 million respectively by 2020. Uzbekistan expects a population of 37.4 million by 2020.

Gross national income per capita in the region as a whole will rise from 681 US$/person to 2260 US$/person. Calculations show that from as early as 2005 daily calorie consumption will exceed 3000 calories per day. This means that there will be an opportunity in the region to produce more food than is required. Here it should be noted that such earlier achievement of rational nutrition norm in these scenarios is explained by high rates of new irrigated land development in all the states. Thus, in the period 2000–2020 there are plans to develop an additional 3360 thousand hectares of irrigated land. The Kyrgyz Republic plans to develop an additional 74 thousand hectares by 2020, while Uzbekistan plans to expand its area of irrigated lands up to 6340 thousand hectares. Turkmenistan and Tajikistan plan to expand irrigated land areas up to 2353 thousand hectares and 1184 thousand hectares respectively. Taking into account the fact that input of new irrigated lands ranges from 5 to 7 thousand US$/ha, and that economic efficiency is to be reached in 13 years (according to expert forecasts), 20 billion US$ will have to be invested over a period of 20 years besides the reconstruction of existing lands. This is more proof of the need for close cooperation between the Aral Sea basin states.

Regional cooperation should be based on the economic benefit from production and must cover areas such as meat and milk production; development of livestock-breeding in Kazakhstan, Kyrgyzstan, and Turkmenistan; production of sugar beet in Kyrgyzstan, early vegetables in Turkmenistan and Uzbekistan, and fruit and melons in Tajikistan, Uzbekistan, and Turkmenistan. The development of hotbed production of early vegetables, and production of canned fruits and dried fruits, will contribute to an increased export potential in the agricultural sector. Regional cooperation will lead to reduced food imports, with small exports inside the region based on the economic feasibility of producing specific crops by zones.

Water requirements in the basin will amount to 149 km^3/year by 2020. Irrigation water requirements will be 132.8 km^3/year. Daily calorie production in the basin as a whole will reach 5081 cal/day/person, resulting in a position of "full food security."

3.3. VISION UNESCO

According to the "Vision" scenario, GNP per capita will be increased 2.5 times by 2025 and will be 1750 US$/person in 2020. "Vision" does not expect considerable growth of the irrigated area, which will be increased by 500,000 hectares over 20 years. The daily nutrition norm will be met through low population increase, with the total population reaching 53 million by 2020, and through high water use efficiency. Daily calorie production in the region will reach 3971 cal/day/person by 2020.

The basic results of the calculations are displayed in Table 2, below.

4. Conclusions

The Aral Sea Basin contains some 41 million people, with 3.5 million of them living in the disaster zone which encompasses most of Karakalpakstan and the Khoresm region of Uzbekistan, the region of Dashowuz in Turkmenistan, and the south-central portion of Kazakhstan. The Central Asian states have the highest population growth rates of the former states of the Soviet Union, averaging around 2.5% a year. This is enough to double the region's population every 28 years. If current growth rates continue, the region's population is forecast to top 60 million by 2020. This will put high demands on food production and water.

TABLE 2. The basic results of calculations according to the offered scenarios for 2020

Scenario	Indicators	Edit	S. Kazakhstan	Kyrgyzstan	Tajikistan	Turkmenistan	Uzbekistan	ASB
2000	Population	Million. people	2.61	2.30	6.12	5.32	24.79	41.02
I			2.98	3.03	7.61	7.51	32.13	53.26
II			3.01	3.06	7.82	7.68	32.98	54.56
III			3.05	3.10	7.77	9.80	36.48	60.19
National Vision			4.81	3.50	8.90	11.68	37.48	66.37
Vision UNESCO			2.98	3.03	7.98	7.07	32.52	53.57
2000	GNP per capita	$/person	1306	637	307	617	729	681
I			3012	1397	908	1883	1610	1615
II			2391	1212	599	1059	1155	1133
III			1895	1048	406	469	766	743
National Vision			3595	1344	715	5490	1738	2394
Vision UNESCO			3341	1549	891	2003	1765	1741
2000	GNP	mlrd.$	3.2	1.5	1.9	3.3	18.1	27.9
I			9.0	4.2	6.9	14.1	51.7	86.0
II			7.2	3.7	4.7	8.1	38.1	61.8
III			5.8	3.2	3.2	4.6	27.9	44.7
National Vision			11.3	4.2	6.2	64.1	65.2	151.0
Vision UNESCO			10.0	4.7	7.1	14.1	57.4	93.3
2000	Contribution into GNP from agriculture	mlrd.$	0.9	0.4	0.4	0.5	5.5	7.7
I			2.9	1.4	1.7	3.3	12.4	21.6
II			2.2	1.1	1.0	1.6	10.4	16.3
III			1.7	0.9	0.6	0.7	8.5	12.4
National Vision			3.6	2.1	1.9	9.9	15.6	33.1
Vision UNESCO			3.2	1.6	1.7	3.3	13.8	23.5

(cont.)

TABLE 2. (*Continued*)

Scenario	Indicators	Edit	S. Kazakhstan	Kyrgyzstan	Tajikistan	Turkmenistan	Uzbekistan	ASB
2000	Irrigated lands	1000 ha	700	416	718	1735	4234	7802
I			863	441	759	1928	4512	8504
II			832	435	747	2045	4392	8451
III			782	402	708	1846	4271	8008
National Vision			815	493	1184	2353	6340	11185
Vision UNESCO			866	442	759	1928	4512	8507
2000	Irrigated lands per capita	Ha/person	0.32	0.18	0.12	0.33	0.17	0.19
I			0.29	0.15	0.10	0.26	0.14	0.16
II			0.28	0.14	0.10	0.27	0.13	0.15
III			0.26	0.13	0.09	0.19	0.12	0.13
National Vision			0.28	0.16	0.14	0.20	0.11	0.18
Vision UNESCO			0.29	0.15	0.10	0.27	0.14	0.16
2000	Total investments in economy	mlrd.$	0.9	0.4	0.6	1.4	2.7	6.0
I			3.3	1.5	2.8	6.4	22.2	36.2
II			2.3	1.1	1.6	3.6	11.1	19.7
III			1.6	0.8	0.9	2.0	4.2	9.4
National Vision			4.2	1.7	3.0	41.7	32.6	80.2
Vision UNESCO			3.7	1.6	2.8	6.4	24.7	39.2
2000	Foreign direct investments	mlrd.$	0.07	0.03	0.02	0.20	0.10	0.42
I			0.53	0.24	0.36	1.34	3.34	5.80
II			0.28	0.13	0.13	0.63	1.03	2.21
III			0.12	0.06	0.03	0.28	0.16	0.66
National Vision			0.67	0.27	0.40	8.75	4.89	14.59
Vision UNESCO			0.59	0.26	0.37	1.34	3.70	6.26

2000	Total water demand by all sectors	km^3	11.8	4.6	12.2	23.3	53.8	105.6
I			9.7	4.3	8.6	20.2	48.3	91.1
II			11.0	4.5	10.1	24.3	52.0	101.8
III			12.0	4.4	11.6	25.1	56.0	109.1
National Vision			18.6	6.1	20.3	27.5	72.8	145.3
Vision UNESCO			9.6	4.4	8.1	15.4	43.8	81.3
2000	Total water demand by irrigation	km^3	10.7	4.3	9.7	22.7	48.6	96.0
I			8.4	3.9	6.8	18.7	42.3	80.1
II			9.7	4.2	8.0	23.3	45.8	90.9
III			10.6	4.1	8.9	24.1	49.0	96.8
National Vision			10.9	6.0	17.4	28.4	70.3	133.0
Vision UNESCO			8.3	3.9	6.2	13.9	37.6	69.9
2000	Total water demand by Industry	km^3	0.229	0.065	0.939	0.074	0.930	2.237
I			0.333	0.150	0.328	0.825	1.653	3.290
II			0.346	0.142	0.374	0.268	1.419	2.549
III			0.414	0.141	0.956	0.104	1.437	3.051
National Vision			0.421	0.150	1.342	3.451	2.263	7.628
Vision UNESCO			0.370	0.167	0.338	0.825	1.833	3.532
2000	Total domestic water demand	km^3	0.122	0.071	0.740	0.411	2.680	4.024
I			0.208	0.212	0.608	0.601	2.763	4.394
II			0.181	0.153	0.786	0.604	3.201	4.925
III			0.152	0.093	0.939	0.757	3.943	5.885
National Vision			0.221	0.205	0.742	0.934	3.138	5.240
Vision UNESCO			0.208	0.212	0.638	0.565	2.796	4.421

(cont.)

TABLE 2. (*Continued*)

Scenario	Indicators	Edit	S. Kazakhstan	Kyrgyzstan	Tajikistan	Turkmenistan	Uzbekistan	ASB
2000	Average daily production of calories	KCal/day/person	3591	2499	1373	3386	2480	2500
I			5911	3424	2094	4750	3468	3587
II			4386	2695	1598	3872	2656	2773
III			2936	1892	1139	2096	1804	1828
National Vision			6567	2341	3914	4943	5510	5081
Vision UNESCO			6093	3428	2250	6696	3658	3971
2000	Amount of food (in calories) need to be imported/exported	mlrd. Kcal	529	−429	−3638	750	−4708	−7497
I			3169	470	−2519	4806	5493	11418
II			1526	−341	−4008	2447	−4143	−4519
III			−71	−1254	−5281	−3239	−15942	−25787
National Vision			4114	−759	2884	8292	33470	48000
Vision UNESCO			3367	474	−2186	9544	7820	19020

TABLE 3. Population figures for the five Central Asian states (including land area and percent irrigated).

State	Population (2000 yr., millions)	Average population growth rate (1990–2000, %/year)	TFR (2000)	Land area (1000 sq. km)	Irrigate land (1000 Ha)
S. Kazakhstan	2.45	1.3	2.3	344	786
Kyrgyzstan	2.34	1.4	3.3	199	422
Tajikistan	6.12	1.6	4.3	141	718
Turkmenistan	5.32	2.8	4.0	488	1735
Uzbekistan	24.79	2	3.8	447	4234
Aral Sea Basin	41.02	1.9	3.5	1619	7895

Assessments of possible future food supply in the region under different scenarios (as detailed above) show that the Aral Sea Basin region has the opportunity to maintain self-sufficiency in food supply until at least the year 2020. Regional available calorie per capita is increasing under the Optimistic scenario and will be more than 3000 Kcal per capita per day (meeting the WHO standard). At the same time, GNP per capita will increase more than 2.5 times in the same period. Such a situation will benefit the whole region. To meet these conditions for each state of the region it is obvious that future socioeconomic development of the region should based on regional economic cooperation. There are obvious agricultural benefits. For example, development of livestock breeding takes place in Kazakhstan, Kyrgyzstan, and Turkmenistan; sugar beet is produced in Kyrgyzstan, early vegetables in Turkmenistan and Uzbekistan, and fruit and melons in Tajikistan, Uzbekistan, and Turkmenistan. Regional economic cooperation would allow these countries to benefit from each other's specialties.

In order to raise the standard of living in the region, irrigated agriculture must be revitalized and related institutions and infrastructures must be optimized. Controlled restructuring of state-farms, the creation of regional markets, and a financial and economic support system are needed to buck the trend of declining production. To achieve this, regional cooperation and agreements are required. This is clear if we examine the Optimistic scenario, where the region is developed through the improvement of integration processes, on the trade of agricultural products, and on the development of regional policy for joint management and efficient use of water.

References

1. Forester, D. J., Machlis, G. E. (1996). Modeling human factors that affect the loss of biodiversity, *Conservation Biology* 10:1253–1263.
2. Indicators of Sustainable Development Framework and Methodologies, UN, 1996.

3. Sustainable Development for Agricultural Sector, UN FAO, 1996.
4. Water Related Vision for the Aral Sea Basin for the Year 2025, UNESCO, 2000.
5. Environment Monographs No 83 OECD Core Set of Indicators for Environment.
6. Kulshreshtha, S. N. (1993). World Water Resources and Regional Vulnerability: Impact of future changes. *International Institute for Applied Systems Analysis*, Laxenburg, Austria.
7. A Vision of Water for Food and Rural Development. World Water Forum. March 2000. Hague, The Netherlands.
8. National Sovereignty and International Watercourses. Green Cross International 2000.
9. Mohinder, S. M. Kyrgyz Republic. Strategy for Rural Growth and Poverty Alleviation. World Bank discussion paper No. 394.
10. Ruziev M.T., Prikhodko V.G. (2002). Assessment of Prospects of Sustainable Development of States in the Aral Sea basin With Help of Model Calculations, *Melioration and water economy* **1**.
11. Avakyan, I. S., Prihodko, V. G., Ruziev, M. T. (2000). “Globesight” Model as a Decision Support System for Aral Sea Basin Water Vision in XXI Century. *SANIIRI Collection, Tashkent.*
12. Dukhovny V. A., Avakyan I. S., Prikhodko V. G., Ruziev, M. T. (2000). The Aral Sea Basin and Irrigated Agriculture in Central Asia in XXI Century. *Melioration and Water Economy* **3**, 2000 (This article was also reported at II World Water Forum, Hague, The Netherlands, March 2000.).
13. Avakyan, I. S., Prihodko, V. G., Ruziev, M. T. (2000) Develop Unified Approaches to Development of Perspective Directions of Strategy of Rational Water and Lands Resources Use With Positions of Vision for “I Century”. *Report SIC ICWC, Tashkent, December.*
14. Uzbekistan, Economic Trends Quarterly Issue, Tashkent, 1999, 2000.
15. Концепция и инструменты Устойчивого Развития, Проект UNDP “Развитие потенциала бассейна Аральского моря”, Ташкент, 1998 г.
16. Иохим Х. Спангенберг. SERI, Научно-исследовательский институт по устойчивому развитию Европы. Определение пространства окружающей среды в регионе для бассейна Аральского моря в Центральной Азии. Материалы № 3 по Центральной Азии, 2000 г.
17. Закон "О воде" Республики Кыргызстан
18. Комплексные основы развития Кыргызской Республики
19. Проект "Национальной стратегии и плана действий по реабилитации ирригационных систем"
20. Указ Президента от 6.10.1997 г. "Об основах внешней политики Кыргызской Республики в области использования водных ресурсов рек, формирующихся в Кыргызстане и вытекающих на территорию сопредельных государств"
21. Водный кодекс Республики Таджикистан, Душанбе 2000 г.
22. Национальная водная стратегия Республики Таджикистан (Душанбе 1995)
23. Среднесрочная программа по выводу из кризиса отраслей агропромышленного комплекса Республики Таджикистан и приоритетные направления стратегии развития его отраслей на период до 2005 г., Душанбе 2000 г.
24. Стратегия развития, основные показатели прогноза социально-экономического развития Республики Таджикистан на 2001–2005 гг. Журнал "Экономика Таджикистана", министерство экономики и внешних экономических связей Республики Таджикистан. Душанбе 2001 г.

25. Программа Президента Туркменистана "Стратегия социально-экономических преобразований Туркменистана на период 2010 года". Ашгабад 1999 г., Министерство экономики и финансов Туркменистана и Национальный институт статистики и прогнозирования.
26. "Программа конкретных действий в зоне Приаралья", 1994 г. Нукус
27. "Предложения по национальной водной стратегии Республики Узбекистан", 1996 г.
28. "Программа конкретных действий в зоне Приаралья", 1994 г. Нукус

ADDRESSING THE NEED FOR INTEGRATED WATER RESOURCES MANAGEMENT AT THE REGIONAL LEVEL

PROSPECTS FOR CENTRAL ASIA DEVELOPMENT—INTEGRATED WATER RESOURCES MANAGEMENT AS REGIONAL ISSUES SOLUTION

V. A. Dukhovny
Director SIC ICWC, B-11, Karasu-4, Tashkent 700187, Uzbekistan

No one needs to be convinced that water is a source and sustainer of life, a guarantee of purity and health, a base of production, an energy producer and one of our most basic needs without which nature cannot survive. Water is one of the fundamental driving forces of sustainable development and it is to the benefit of humanity that it is a renewable resource. However, we are not always its wise consumers and users. Despite multipurpose water use and its multilateral involvement in all spheres of human activity, it has traditionally been perceived in a narrow and specialized manner, leading to dispersed management of its different functions, isolated actions without consideration of other uses for water and frequent misunderstanding of the complex impact and consequences of water use. An understanding of water's importance came to humanity in the last quarter of the twentieth century, and even then not universally but to certain groups and users. But the road from understanding to action is a long one, with many obstacles and destabilizing factors along the way.

In Central Asia, it is impossible to overstate water's importance and meaning, not least because irrigated farming is the basis of agriculture in the region. Nevertheless, regarding the political climate, water has lost its position of prime importance. Water issues are solved in ministries of the environment or agriculture, ministries of agriculture and water resources or processing industries. This situation is not accidental—within society there is understanding of current issues. Even in the most water-abundant zones only 50% of the population is provided with sustainable water supply and even less by adequate sanitation facilities. Irrigation systems in Kazakhstan, Turkmenistan and Uzbekistan, with their thousands of kilometers of canals, have fallen into a state of disorder and all the governments are trying to maintain them. The same problems accumulated in irrigated farming—huge funds created in the past are exhausted and the situation is only maintained due to the enthusiasm of operational staff exploiting their knowledge and experience.

Another catastrophe is imminent. Human resources are also wearing thin, with only small measures taken to support and develop potential candidates

P. Wouters et al. (eds.), Implementing Integrated Water Resources Management in Central Asia, 127–138.

who can replace existing specialists. Budgets and loans in all of these countries are devoted to supporting significant efforts in this sector but they do not take into consideration future development, including human potential development and support.

Transition in the marketplace also impacts upon the water sector. The market is oriented towards quick results, quick revenue, and profit making, and pays less attention to things that can give good results in the future. But all water-related measures have long-term characteristics. They are conservative in nature and their long-term development objectives come into conflict with the market's demands for instant profit. It is not coincidental that all international finance institutions have reduced their investments in the water sector. Moreover, the market is oriented towards narrow interests in order to make the maximum profit. But who will pay for the poor man, or for nature itself? If current national approaches and self-interests stand in the way of human water development, a range of external factors also work against us:

- population growth, whose well-being depends on direct or indirect use of land and water resources;
- six different states within the Aral sea basin, each with their own national and sectoral interests and different levels of economic maturity and development;
- globalization expressed in both evident and hidden impacts: low prices on raw materials including food; fight for markets and import prices dumping, attempts of large companies to take control over water and power resources and dictate their own "rules of the game" in transition of the economy;
- climate change, resulting in extreme natural phenomena (flood, drought), increasing in frequency and severity.

Under these conditions, we must take a sensible approach and "look forward" in the water sector to consider what we might set against objective and subjective trends and conditions. We must consider what actions can be taken in order not to lose the achievements of the last 10 years. We must heed the Eastern proverb that says: "the path can be ploughed through by walking," and the Japanese saying: "Climb easier together."

In order to perceive how we should conduct ourselves and make progress in this situation, we should take the position of those whom we serve and whom we want to (and must) make our allies, assistants and partners—the water users.

Consider the figures below, relating to average water availability in Ferghana valley canals for 2003, and the even stronger dispersion of it within WUAs, and think how crop-growing technology could be implemented under such water supply fluctuations.

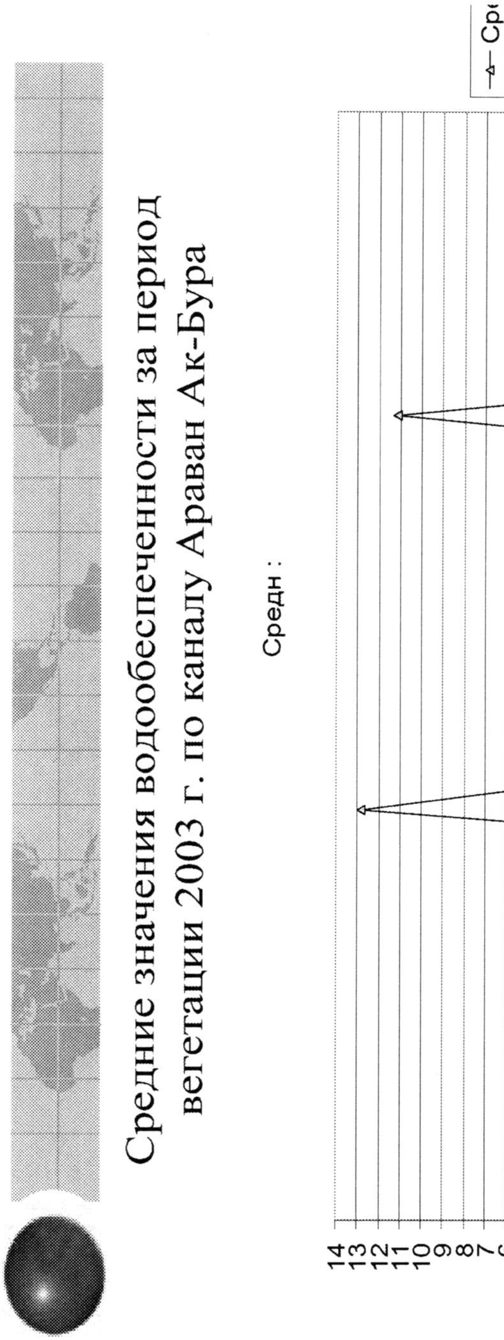
Средние значения водообеспеченности за период вегетации 2003 г. по каналу Араван Ак-Бура
Средн :
Сре
0
1
2
3
4
5
6
7
8
9
10
11
12
13
14
Ак-Бура
Ак-Бура
Ак-Бура
Мамакеев (?) (отвод)
Чон-Сай (?) (отвод)
Шамши (отвод)
Мектеб (отвод)
Шыйлан (отвод)
Баня 1 (отвод)
Баня 2 (отвод)
Толонбай (отвод)
Сатар лев. (отвод)
Сатар прав. (отвод)
Маяковский (отвод)
Толпу (отвод)
Рахматилла (отвод)
Аман (отвод)
Каирма (голова) (отвод)
Янги (?) (отвод)
Санаторий дет. (?) (отвод)
Родром (отвод)
Аскар (отвод)
5-петка верхний (отвод)
5-петка нижний (отвод)
Каримова (отвод)
Ката-Хас1 (отвод)
Сельпо (отвод)
Ката-Хас2 (отвод)
Шабдан (Жалалак) (отвод)
ПМК (отвод)
Труба (отвод)
Абай (отвод)
Анарбай лев (отвод)
Анарбай прав.(орке) (отвод)
Насир(Кок-Добо) (отвод)
Тутовник (Бак) (отвод)
Трансформатор (отвод)
Таштан (отвод)
Новый (отвод)
Лоток (отвод)
Сброс 1 (отвод)
Отвод 1 (отвод)
Отвод 2 (отвод)
Отвод 3 (отвод)
Отвод 4 (отвод)
Отвод 5 (отвод)
Сброс 2 (отвод)
Новый отвод (отвод)
Сброс 3 (отвод)
Дюкер (отвод)
Партсъезд (отвод)
Кирпичный (отвод)
Коммунизм (отвод)

Средние значения водообеспеченности за период вегетации 2003 г. по каналу Ходжа-Бакерган

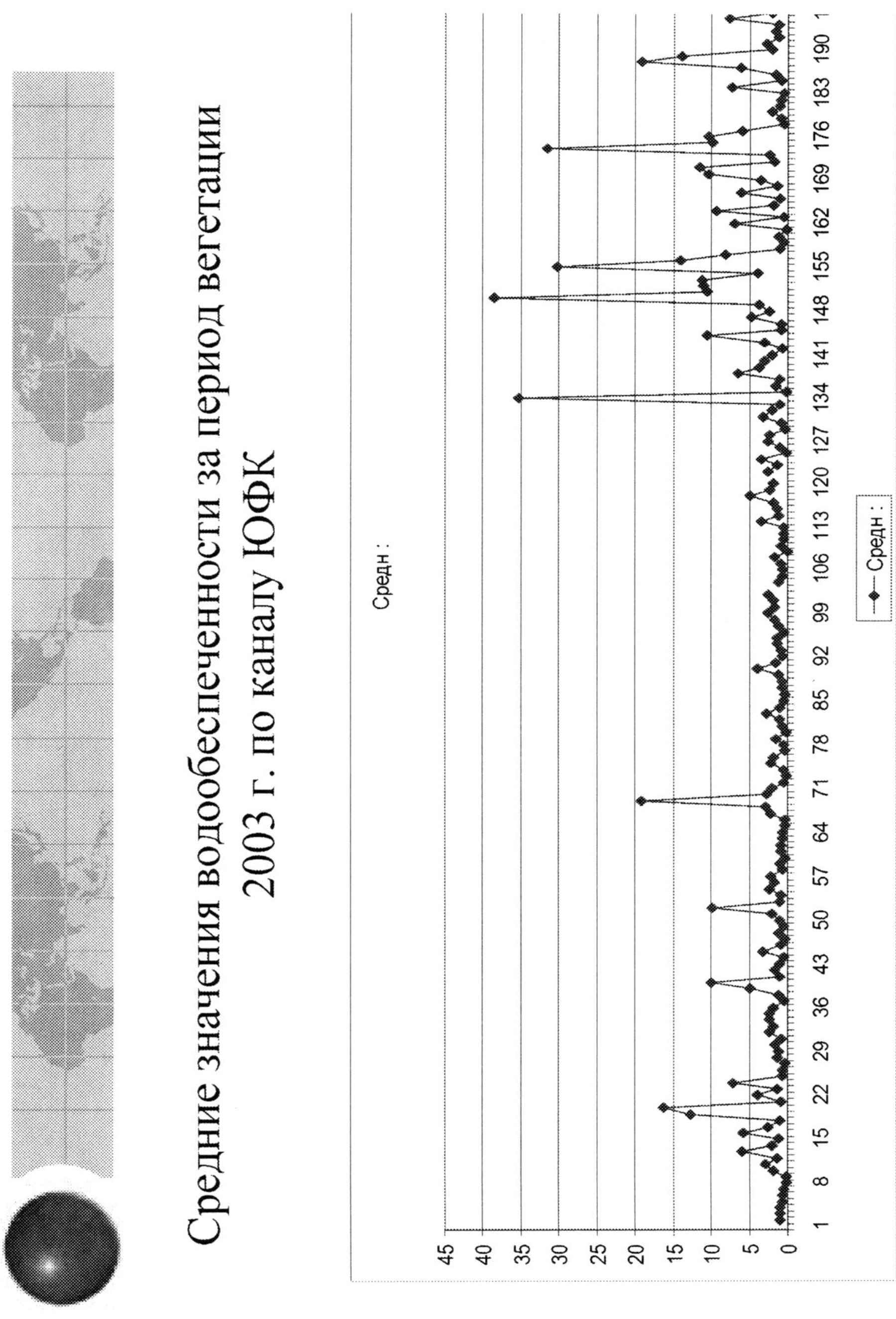
Средние значения водообеспеченности за период вегетации
2003 г. по каналу ЮФК
Средн :
Средн :
0
5
10
15
20
25
30
35
40
45
1
8
15
22
29
36
43
50
57
64
71
78
85
92
99
106
113
120
127
134
141
148
155
162
169
176
183
190

Result of pilot canals management during growing season 2003

No.	Indicators	Unit	AABC	Hoja-bakirgan	SFC
	Specific water supply per 1 ha	th. m^3	12.2	14.00	10.53
1	Water availability coefficient in branches' head:				
	minimum =		0.15	0.17	0.44
	mean =		0.63	0.81	1.10
	maximum =		1.19	1.02	1.97
2	**Operation efficienty:**				
	minimum =		0.71	0.44	0.88
	mean =		0.89	0.75	0.96
	maximum =		1.12	0.87	1.03
3	**Ratio between total diversion and evatranspiration**		1.61	1.81	1.45

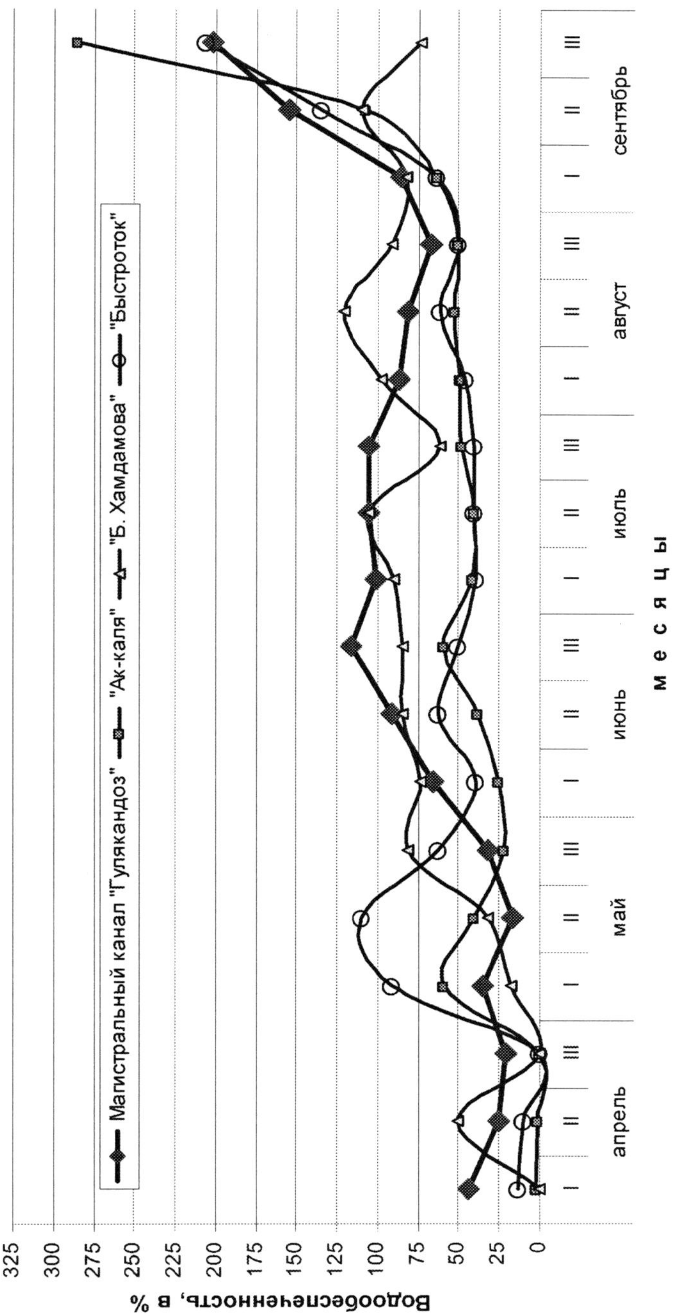

Рис. 2 Динамика водообеспеченности каналов АВП "Заравшан" из канала "Гулякандоз" за вегетационный период 2003 года

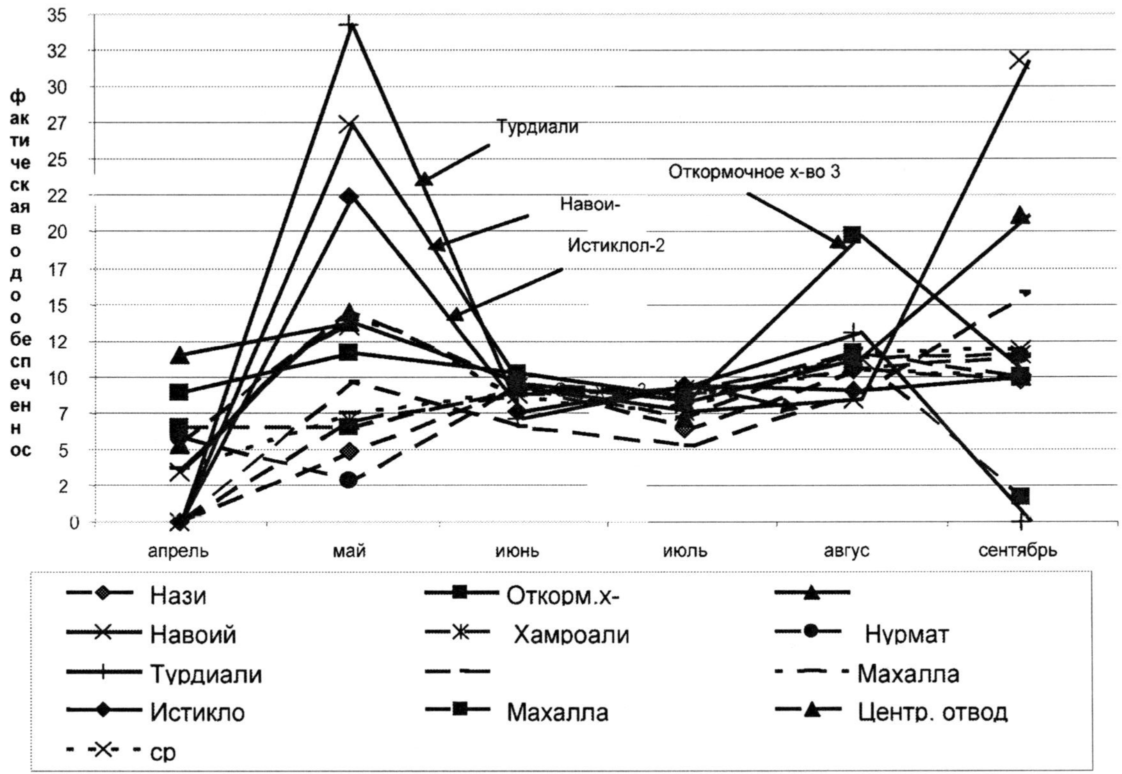

Фактическая водообеспеченность водопользователей, подвешенных к каналу "Акбарабад-2", за вегетационный период 2003 года,

It is clear that the old "top–down" water supply system must be replaced by stochastic water expectation through many staged and multifactor staircases from the basin to the user. Where there are so many willing and seeking to intervene it is sometimes surprising that the consumer obtains water at all. Indeed, sometimes it seems that this occurs by chance.

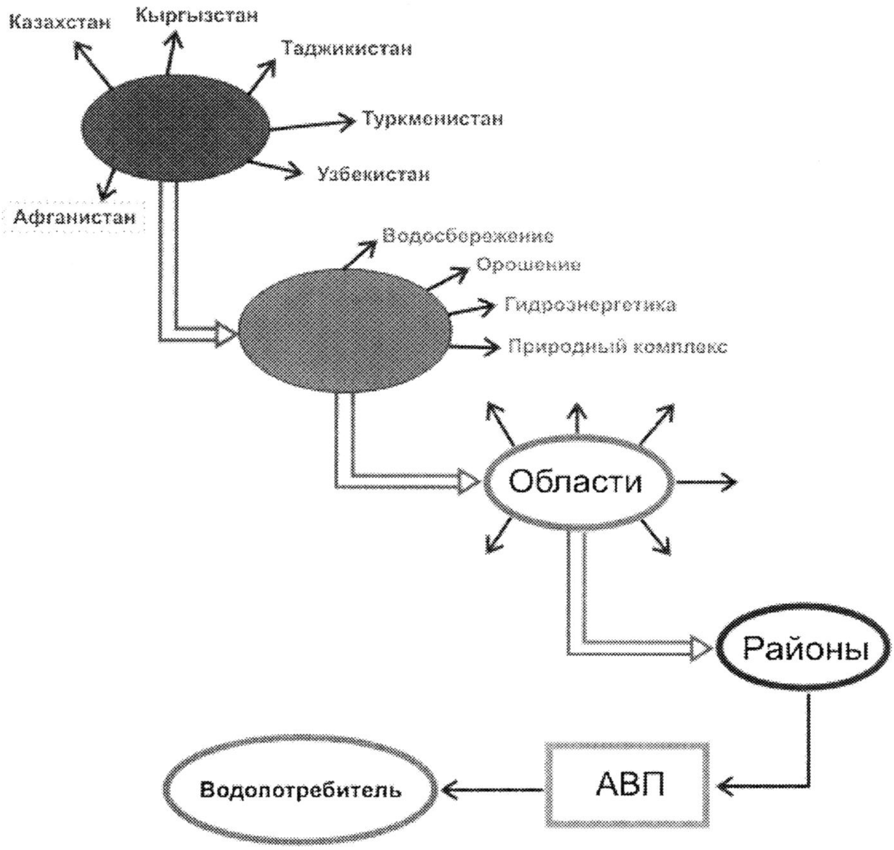

Integration is the single key to the solving of the very complicated issue of water management and preservation and this solution should be implemented from a higher level and coordinated between all hierarchic levels of water management. Integration, which we started to become aware of in the example of three pilot canals in the Ferghana valley, and which is described in detail in V. Sokolov's recent paper, is only a part of the process, related to lower hierarchic levels like the canal, WUA, and water user. I would like to draw your attention to issues more of principal such as relations between governance, society and management.

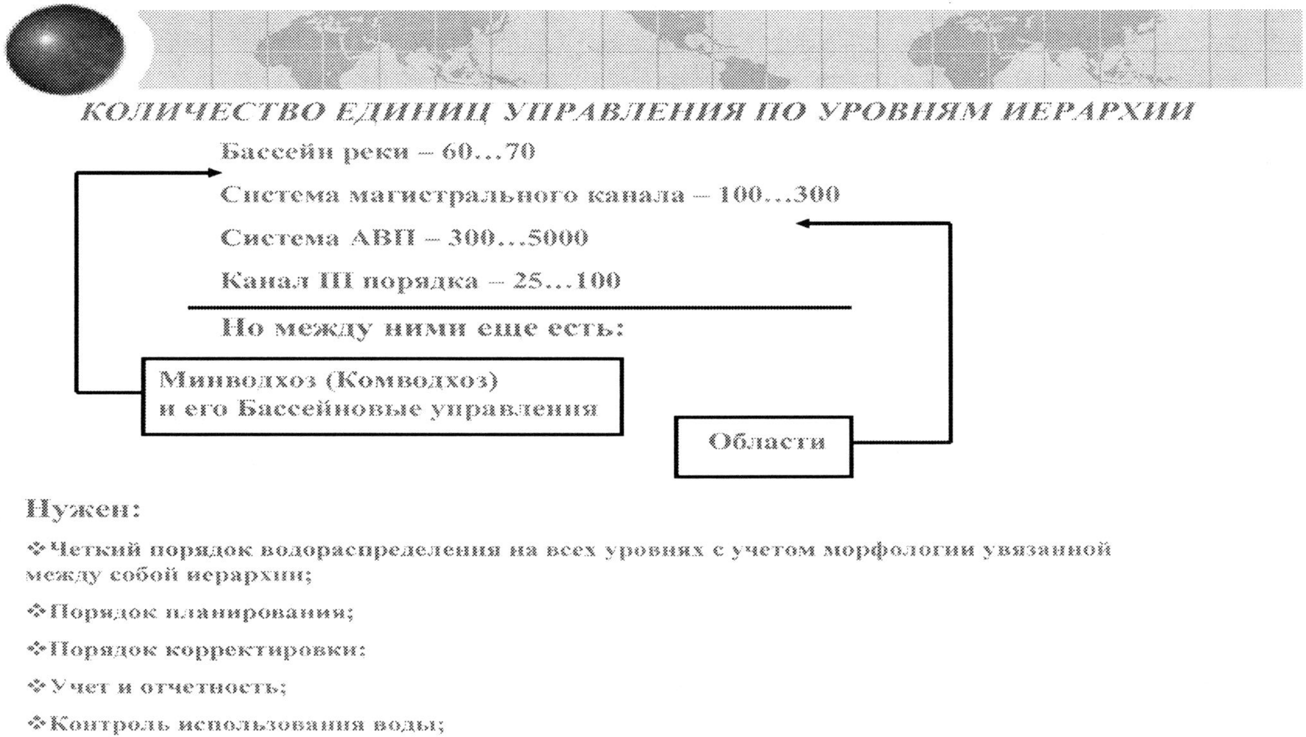
КОЛИЧЕСТВО ЕДИНИЦ УПРАВЛЕНИЯ ПО УРОВНЯМ ИЕРАРХИИ
Бассейн реки – 60…70
Система магистрального канала – 100…300
Система АВП – 300…5000
Канал III порядка – 25…100
Но между ними еще есть:
Минводхоз (Комводхоз)
и его Бассейновые управления
Области
Нужен:
❖Четкий порядок водораспределения на всех уровнях с учетом морфологии увязанной между собой иерархии;
❖Порядок планирования;
❖Порядок корректировки;
❖Учет и отчетность;
❖Контроль использования воды;

What is water governance? It is the relationship of the political sphere of the country to water. This covers things such as the "regulation of property, belonging, distribution, management, public participation, water use and protection as principal source of life" (R. Petrelli). The political sphere determines:

- attitudes to international water rights;
- readiness to open fields of (mutually beneficial) cooperation;
- transparency and awareness;
- water users involvement;
- recognition of IWRM as a key tool;
- promotion of a respectful attitude to water;
- water conservation policy;
- willingness and readiness to invest in the water sector and to support it;
- attitude to environmental requirements.

It would be constructive if each of us who are responsible for the stability of water management at the higher levels could honestly define our activity according to the scale of the above criteria and examine what we do to improve water governance.

An independent survey of unofficial concerned objects completed by GWP in January 2004 produced a very interesting assessment that coincides with my own vision for our situation. The survey was carried out in countries in transition and in developing countries. These countries are divided into groups A, B and C.

- Kazakhstan and Kyrgyzstan are both placed in group A; Kazakhstan is marked very highly ("the best"). This assessment coincides with our criteria. Kazakhstan joined in the International Agreements and Conventions of 1992 and 1997, the society is open and informed, IWRM is recognized as the main direction in water activity at national legislation level and big investments have been made in water sector, including deltas.
- Tajikistan, Turkmenistan and Uzbekistan are in group B, but the most positive situation is found in Tajikistan where "political commitments taken by the government in 2003 gave a good start to processes." In Uzbekistan, IWRM introduction is underway. As to Turkmenistan, the potential for its introduction is limited.

What must we do, then, at interstate level in order to accelerate our movement towards IWRM? We need, ultimately, to arrive at some certain principles of governance and management. We need to decide whether we want to push the process ahead or whether we will continue to play the role of closed clubs of water bureaucrats. If it is the former, certain steps are necessary:

- Integration of all water-related bodies: hydromet service, ground water service, hydropower engineering, environment and municipal bodies. We

cannot work without each other but are afraid to open ourselves and move towards establishing a single platform on water accounting and forecasting by quantity and quality.
- ICWC and BWO should involve in their activities all concerned parties and make them partners rather than external observers.

Recognizing ecological requirements does not necessarily mean water supply for the environment. Kazakhstan really protects its deltas but still plans water supply to the deltas residually.

The agreements of 1998 have run their term and now perform badly. There is, however, a solution. Firstly, both water and power specialists should plan water and energy conservation measures, after which it will be easy to coordinate irrigation schedules (with 15% water savings) and power schedules (with 20% energy savings). Secondly, it is necessary to separate water from fuel. The river regime should be strictly observed; energy and fuel should be coordinated within the water-power consortium as regards the financial mechanism of this exchange. This is important in conditions where the market price of 1 kvt/h in the region does not exceed 1 cent/1 kvt/h. Flow regulation should be paid but not 1 m^3 of water should be released during seasonal or long-term regulation.

There are several other considerations:

- Water resources integration. We still try to jointly manage only surface water but there are huge resources of ground and return water. Special attention must be paid to these waters because they also can be a source of conflict. In their agreements, Israel, Jordan and Palestine take into account each million of cubic meters of these waters, whereas we ignore tens of billions Of cubic meters
- Water distribution accuracy and confidence improvement through head structures equipment by SCADA systems. The total cost of this work does not exceed 30 million USD. From this, we could define who takes water, and in what quantity.
- Integration of interests is based on aspiration to satisfy and understand the requirements of others. Upper reaches require participation in terms of their expenses for watershed maintenance and lower reaches require maintenance of quality and ecologic releases and payment for river regime violation. All this is fair. These unsolved issues make our management unsustainable and subject to risk. It is impossible to opt out of these questions and it is necessary to organize routine activity of all countries representatives and organizations to perfect systems of water governance and management. We should look ahead. If we do not do this we will face much greater expense and trouble in the future.

INTERSTATE, INTER-SECTORAL SCIENTIFIC AND PRACTICAL INTEGRATION

Abdukohir A. Nazirov
Minister of Land Reclamation and Water Management,
The Republic of Tajikistan

With the establishment of the independent states on the territory of Central Asia, water allocation issues have undergone a significant change, moving from the internal affairs category to become an interstate problem of an economic and political nature. It should be added that water resources use issues acquire acute urgency due to population growth in the region and an increase in anthropogenic pressure on the environment. The Republic of Tajikistan, being a Party to the Agreement between Central Asian states on cooperation in the field of water resources use concluded in Almaty in 1992, advocates their further development and harmonisation.

The order of setting quotas for water resources use (which is a remnant of the former Soviet regime) exists in the region against the background of non-availability of the economic mechanism that had been previously functioning in the field of interstate water use. This leads to the deterioration of significant, interstate waterworks facilities and water reservoirs and, even more critically, to losses of scarce water resources. Apart from this, failure to fully comply with annual intergovernmental agreements on water and energy resources use compounds the problem. Unfortunately, this proves to be true even as applied to the Syrdarya river basin, despite the fact that this basin has been perfectly well regulated.

Due to unadjusted economic interrelations, regulating capacity in this basin is not sufficiently utilised; emergency situations occur which can sometimes lead even to the brink of catastrophes. This was demonstrated by the experience of operating Arnasay and Chardara reservoirs during the vegetation period of 2000 and the winter-spring period of 2003. These events showed that water resources use requires the taking into account of not only short and medium-term measures but of long-term factors as well.

The time is now ripe for transition from low-effect, annual bilateral agreements (between, for example, Tajikistan and Uzbekistan and Uzbekistan and Kyrgyzstan) to long-term multilateral agreements. Our countries are located in a single integrated basin, and the manner in which we have been addressing the challenges at present is clearly not an example of a fully integrated water resources management system.

P. Wouters et al. (eds.), Implementing Integrated Water Resources Management in Central Asia, 139–145.

Kayrakkum water reservoir, which is subject to seasonal regulation and has a full capacity of 4.6 km^3, is the central element in the system of the Naryn-Syrdarya cascade of reservoirs. Kayrakkum reservoir plays a significant role in providing irrigation water supply to Uzbekistan, Kazakhstan and Tajikistan. Water management agencies from the interested parties have developed proposals on reconstruction of the Kayrakkum hydro-scheme that involve augmenting the capacity of this reservoir. To this end, a certain amount of construction work is required, enlarging the dams' height, removing silt from the reservoir floor, modifying the HPS gates and reconstructing the drainage systems and pumping stations. To maintain the operational water level in the reservoir (347.5 m), immediate restoration of dams is required amounting to a total length of 8,000 m. Unless all of these measures are implemented, we will be forced to sustain the water level at a mark, which will, from next year, be lower by 1 m (346.5).

The current practice of water use in the Syrdarya river basin does not take into account the interests of Tajikistan to a full extent. Expenses connected with the operation of Kayrakkum reservoir far exceed the amount of compensation provided by downstream water users. There is a need for elaboration of amendments to the current 1998 Agreement on water-energy resources use in the Syrdarya basin, in terms of determining a better-defined economic mechanism of water and energy resources use, as well as developing, on the basis of such a revised Agreement, the system of unconditional compliance with annual contracts concluded between governments and the economic entities which represent contracting parties. The project of augmenting Kayrakkum reservoir capacity should include a special section dedicated to international use of this water facility. With the support of international donors, a working team of experts from interested countries could start considering the issue immediately.

The most serious attention should be paid to the Amudarya river basin, which is insufficiently regulated, as well as to the possibility of substituting the regulating Nurek and Rogun reservoirs, whose construction is incomplete, with some other reservoir, for example, the Dashtidjum reservoir, which has a capacity of 17.5 billion m^3 and energy output of 5.2 million kWh. The total possible volume of regulated run-off in Tajikistan constitutes approximately 68 km^3/year; this is equal to 56% of average long-term run-off in the Aral Sea basin.

The completion of the Rogun hydro-scheme construction will allow greatly improved reliability of irrigation for agricultural purposes in the middle and lower reaches of the Amudarya. The clean electrical energy which is produced concurrently may be delivered to the regional market. The only precondition for this is a willingness on the part of neighbouring countries to participate in such a scheme. The Regional Advisory Commission of the SPECA Programme has acknowledged the Rogun hydro-scheme as a facility of regional significance, and has suggested it be considered a priority for investment. To construct the reservoir, a total of 120 million USD are needed. It will cost 300 million USD to put the first generating unit into operation, and full completion would cost a sum of 800 million USD.

The Nurek hydro-scheme on the Vakhsh River is a facility for seasonal water resources regulation in the Amudarya river basin. During the former Soviet period, 'accompanying' electric energy generated in summer by this HPS had been delivered to consumers in Central Asian republics. In exchange for this, Tajikistan received electrical energy and vital fuel-energy resources in winter. Since then, the situation has changed radically—There is no market for summer energy generated by Nurek HPS and water is provided to neighbours, whereas in winter Tajikistan buys electric energy at a price two times higher. Therefore, there is a need for cooperation and a mutually acceptable trade-off regarding decisions on improving the efficiency of water and energy resources use in the region. It is necessary to start drafting an Interstate Agreement on water and energy resources use in the Amudarya river basin, at least at the level of a framework document similar to that on the Naryn-Syrdarya cascade signed in 1998. This would allow the efficient use of renewable water-energy resources and the achievement of clearly positive ecological effects in contrast to those produced by thermal power stations.

Great wisdom, and a truly regional approach, was demonstrated by the heads of the Central Asian states in decisions taken within the framework of the Organization for Central Asian Cooperation, with regards to establishing water-energy, transportation and food consortia. The creation of a water-energy consortium is a step in the right direction towards Integrated Water Resources Management. The task we must fulfil is to implement these decisions unconditionally. We must work on corresponding feasibility studies, for which we need the necessary technical assistance on the part of international financial organisations. We must then find ways of implementing these decisions.

Integration, cooperation and specialisation—This is what we need now. Then we will have sufficient water, energy and food, and the sustainable development of the states of Central Asia will be ensured.

The Decisions taken by Heads of Central Asian states in Dushanbe on October 5–6, 2002, substantiated the idea of establishing a special UN commission for the co-ordination of assistance activities carried out in the region, as well as imparting the status of a UN structure to IFAS. It is necessary for the regional organisations (ICWC, SIC ICWC, BWOs 'Syrdarya' and 'Amudarya') to become, in essence, international organisations—this will also facilitate the introduction of Integrated Water Resources Management in the region.

Specialists from Central Asian states have developed a number of specific projects included in a draft ASBP-2 Program with respect to improving the socio-economic and ecological situation in the Aral Sea basin. The presentation of the ASBP-2 took place during the recent Dushanbe International Fresh Water Forum.

A series of projects on the practical introduction of Integrated Water Resources Management approaches in the region are in the process of being implemented. A good example is provided by the project on 'Integrated Water Resources Management in Ferghana Valley'. Within the framework of this project, a number of pilot fields are operated where various approaches are tested with respect to introducing hydro-graphic management principles, stakeholders' participation, water saving technologies at various levels of the resource use, and inter-sector coordination. Introduction of an integrated method of water resources management is one of the ways to achieve the millennium goals adopted in Johannesburg with regard to sustainable development problems.

At present, the following issues take on special significance: (i) effective use of investments; (ii) capacity building of water management institutional structures; (iii) coordinated interaction of donors, international, governmental and non-governmental organisations; (iv) the deepening and extending of cooperation between all parties involved.

Regarding the latter point, we are interested in deepening and extending cooperation along the following lines:

- Attraction of foreign investments for new irrigated land development and improvement of current land use through the application of new technologies;
- Rehabilitation of irrigation and collector drainage systems,—beginning with the most vulnerable structures and those located in the pumping irrigation zone;
- Rehabilitation and development of good quality water supply systems for the rural population (about 50% of rural population use drinking water from adverse open sources);
- Production of water-metering equipment;
- Development and implementation of measures designed to control floods, mudflows and landslides;
- Development and introduction of a long-term programme of natural disaster prevention in riverside zones, systematisation of bank-protecting structures construction and rehabilitation of landscapes in upper water catchment areas.

There is a need for support in the form of grants allocated along the following lines:

- Research and introduction of a differential charge for water use, delivery depending on specific conditions of economic activities, and on development of various forms of property rights in water use (private, collective and joint-stock ownership);
- Training and retraining of specialists able to work in market conditions, developing skills for application of new technologies and establishment of training centres for this purpose;
- Post-privatisation support to farmers and water user associations;
- Rehabilitation of the water use accounting system;
- Creation of a state-of-the-art information and communication system of water management;
- Rendering support in establishing markets for water services;
- Rationalisation of structures and functions with respect to water resources use and protection management;
- Development of policies and strategies in the field of water resources use taking into account national interests of Tajikistan in compliance with International Law.

The achievement of these goals should be facilitated by the appropriate scientific support, and by the exchange of innovative, low-cost water and energy saving technologies designed to provide high crop-yields and profitability in farming. Therefore, experience obtained by developed countries is of great value, and the sharing of such technical assistance is vital to the successful development of such a system in Central Asia.

To enhance integration there are necessary legislative preconditions in Tajikistan. But the key issue is the provision of post-privatisation support to water users, as well as financial support to allow irrigation systems to be brought up to current market requirements. The government of the country is actively attracting investment in the water sector. At present, a number of projects are being implemented on the rehabilitation of irrigation systems, supported by international financing organisations (WB, ADB, USAID and others). Unfortunately, however, these programmes do not cover all of the irrigated area in the country.

It should be noted that the addressing of national issues of water resources management must be regarded as a component of regional objectives; measures directed at extending cooperation should be undertaken as they apply to every hydro-graphic unit. For example, there are important projects being implemented covering trans-boundary water issues in the Ferghana Valley (Large Ferghana Canal and Northern Ferghana Canal).

The major part of Tajikistan is located in the Aral Sea basin flow formation zone. In these areas, the local population is vulnerable to the adverse impacts of mudflows, floods, landslides and other natural phenomena that cause significant tangible damage. Expenses incurred by the Ministry of Water Management for recovery, bank-protecting and river channel regulation works constituted 1 million USD in 2003 and in general, by the republic as a whole, more than 24 million USD. It should be underlined that this problem is of a regional nature, and is of prime importance to Tajikistan among the many issues of regional cooperation in Central Asia. The radical way to combat floods and draughts is to construct, in the flow formation zone, large water reservoirs designed to protect the population and facilities in downstream areas.

Tajikistan possesses hydro-energy resources that constitute about 527 kWh/year, or 4% of the global reserves. Due to a total lack of mineral fuel, the priority in water resources use is given to hydro-energy, a fact which has regional significance. The Republic of Tajikistan invites all interested parties to participate in developing these resources on a mutually beneficial basis, and this call is addressed first of all to the neighbouring countries.

Forecasting water availability is the concern of everyone in the region, and is especially important for low-water years. The major resources of ice and snow are concentrated in Tajikistan and Kyrgyzstan. Therefore, implementation of appropriate joint projects plays an important role in this sphere of activities. The regional water system is interlinked; it is of great economic and life-sustaining significance. That is why the development and implementation of projects on monitoring sources of water are necessary, including the observation of glaciers and mountain lakes. There is a pressing need for the establishment of a regional Hydro-Meteorological Centre, which will serve the purpose of integrating efforts in this direction. Then, it is necessary to develop, within a separate task-oriented project supported by international organisations, new mechanisms of interstate water allocation, with an allowance made for the Aral Sea and for the sustaining of a stable ecological situation in the basin proceeding from new political and economic realities. It is also expedient to develop, within a separate project, an economic mechanism for water use in the region. Such a mechanism should be introduced in a step-by-step manner; otherwise there is no chance of avoiding deterioration in the technical state of the water sector. These requirements are met in full measure by provisions in the 'Strategy for Regional Cooperation in Efficient and Effective Use of Water-Energy Resources in Central Asia' now being developed within the SPECA programme (UN EEC and ESCAP), which is to be submitted for consideration by Heads of Central Asian states by the end of 2003. It is imperative that the continuation of the SPECA programme is guaranteed. In general, these issues were stipulated in a draft programme for the improvement of the socio-economic situation in the Aral Sea basin (ASBP), initiated by the Heads of Central Asian States in October 2002, and should therefore be supported.

Integration is a very comprehensive and holistic concept; it covers various spheres of interaction including single water users, basin management, relations between states, and interstate bodies, as well as covering issues of specialisation, cooperation, funding, private business, information technologies, foreign and internal policies and many other aspects. Step-by-step harmonisation of relations at all levels of natural (and, first of all, water) resources management hierarchy will serve as a sound basis for sustainability and security in the Central Asian region.

THE ROLE OF STRATEGIC AND NATIONAL PLANNING IN THE DEVELOPMENT OF WATER MANAGEMENT

T. Altyev
Deputy Minister of Water Management of Turkmenistan, Turkmenistan

It is difficult to overstate the significance of water management to the economic development of Turkmenistan, particularly in the case of irrigated agriculture, given the location of the country and its arid climate, where the hot drought season lasts between 3 and 5 months of the year, and where evaporation exceeds the amount of natural precipitation by 8–10 times. Because of this, a distinctive attitude towards water has been formed during the Turkmen nation's centuries-long history—That of solicitude, frugal water use, protection of the purity of water sources, and equitable water distribution between users. The *Mirab*—the local water manager—was historically chosen on the basis of democratic elections among fair and honest people, and that is why an opinion expressed by him was perceived as the Law, and was subject to implicit obedience.

Unfortunately, many wholesome traditions of our ancestors have been lost during the extensive irrigated agriculture development in the Aral Sea Basin in general, and in Turkmenistan in particular. The pursuit of ever-growing yields of cotton "at any cost," without adequate engineering surveys and the preparation of irrigation canals and lands under crops have resulted in the large-scale degradation of fertile lands and the occurrence of the Aral Sea problem.

Nevertheless, over the last few decades a high-capacity water management system has been developed in Turkmenistan that provides all sectors of the economy with water, including an agricultural production area of about 2 million hectares. Table 1 (below) shows the technical characteristics of these water sector facilities.

Turkmenistan's water resources (during years of average water availability) constitute approximately 25 billion m^3 and are made up of the Amudarya, Murgap, Tedjen, and Atrek Rivers' surface flows, small watercourses, and underground water. In addition, there are mineralized discharges of collector-drainage (return) water. Table 2 (below) shows available water resources in Turkmenistan.

At present there are three large water management divisions:

P. Wouters et al. (eds.), Implementing Integrated Water Resources Management in Central Asia, 147–156.

TABLE 1. Technical characteristics of land reclamation systems in Turkmenistan (as of 2002)

Description	Units	Total responsibility of Ministry of water management
Irrigated area, total	ha	18,35,000
Demand (withdrawal)	million m^3	22,500.2
Spread of irrigation network	km	42,324.9
including:		
(a) inter-farm	km	7,917.4
with capacity up to 2	m^3/sec	1,039.3
from 2 up to 5	m^3/sec	2,137.4
from 5 up to 10	m^3/sec	2,159.1
from 10 up to 25	m^3/sec	1,315.4
from 25 up to 50	m^3/sec	793.9
from 50 up to 100	m^3/sec	126.1
above 100	m^3/sec	346.2
(b) on-farm	km	34,407.5
Spread of CDS, total	km	37,543.6
including:	km	
(a) inter-farm	km	9,494.6
With depth—down to 1.5	m	334.4
from 1.5—down 2.5	m	2,747.9
exceeding 2.5	m	4,311.4
(b) on-farm	km	27,380.6
out of which	km	
underground drainage	km	10,410.4
(c) interstate collectors	km	668.4
Lined canals	km	3,421.1
including:	km	
inter-farm	km	907.7
out of which	km	
(a) lined with concrete	km	548
(b) ferroconcrete flumes	km	343
on-farm	km	2,513.4
out of which	km	
(a) lined with concrete	km	923.8
(b) ferroconcrete flumes	km	1,589.6
pipelines	km	2,696.3
including:	km	
inter-farm	km	471.4
on-farm	km	2,224.9
Dams	km	1,404.6
Hydro-technical facilities in irrigation network	one facility	11,167
including:		
inter-farm	one facility	2,025
on-farm	one facility	9,142

TABLE 1. (*continued*)

Description	Units	Total responsibility of Ministry of water management
Hydro-technical facilities on CDS including:	one facility	4,871
inter-farm	one facility	1,141
on-farm	one facility	3,730
Hydro-posts	one facility	2,177
Water reservoirs	one facility	14
Total volume	million m^3	3,018
including:	million m^3	
Zeid	million m^3	1,000
Hauzhan	million m^3	700
Kopet-Dag	million m^3	540
The Murgab River water reservoir	million m^3	475
The Tedjen River water reservoir	million m^3	218
The Atrek River water reservoir	million m^3	37
Intakes from inter-farm network		969
Outlets for farms	one facility	5,853
Bridges and crossings	one facility	6,142
Pumping stations	one facility	2,928
Command area	ha	8,36,384
Inter-farm pumping stations	one facility	284
Pumping units	one facility	1,314
Command area	ha	2,82,579
On-farm pumping stations	one facility	2,651
Pumping units	one facility	3,140
Command area	ha	5,53,805
Communication lines	km	888.0
Cables	km	2,329.6
Roads	km	566.5
Out of which	km	
paved	km	357.2
gravel	km	193.4
earth	km	15.9
wells	one facility	1,108
forestation	ha	12.7
Operational personnel	person	4,252
with higher education	person	631

First, there is the Amudarya downstream division, within the administrative borders of Dashgouz *veloyat*[1]. The hydrographic network in this division is represented by large irrigation systems of interstate significance at Gazvat,

[1] *Veloyat* is an administrative province in Turkmenistan.

TABLE 2. Water resources in Turkmenistan (in a year of 50% availability)

River, hydro-post	Flow volumes, mln. m^3	Notes
Amudarya, Kerky	22,000	For Turkmenistan, according to the international agreement
Murgap, Tagtabazar	1,550	
Tedjen, Pulikhatum	770	
Atrek, Chut	170	
Small rivers	310	
Underground sources	270	
Return waters	5,350	Statistic data as of 2002 г.

Shavat, Klychbay, Djumabaysak and Kipchakbozsu. In connection with construction of the Tukmenderya system, the interstate networks of the Gazavat and Shavat systems have been combined into a single system, called "Turkmenderya." Future development of the "Turkmenderya" system also stipulates integrating the interstate Klychbbay and Kipchakbozsu canals into this system, thus ensuring water supply to command areas and to a major part of the Amudarya downstream water management district independent of interstate systems. The Hanyab Canal intakes water from Turkmen section of the Amudarya River.

Next is the Collector district. The drainage network of this water management district is also represented by two large interstate systems—the Ozerny and the Daryalyk, which are not only inter-farm drainage collectors for the district, but also play the role of large main collectors for conveying drainage water from neighboring provinces of the Republic of Uzbekistan. Lately, in operating these large interstate collector-drainage systems, serious socioeconomic and ecological problems have emerged which are to be solved in the context of perspective development of Turkmenistan's economy in general, and water resources development in particular, stipulated for the period until 2020.

The hydrographic network of the Amudarya middle stream water management district is represented by numerous irrigation canals of inter-etrap[2], etrap and local significance, that take water from the Amudarya river through non-barrage intakes, with the exception of the Shih-Bytyk and Bossaga-Kerky Canals, which intake water from Amubukhar and Karakuderya Canals respectively. The collector-drainage network Ко ллекторно-дренажная сеть on the Amudarya left-bank territory is integrated into the system of the Main Left Bank collector, whilst on the Amudarya right-bank territory, independent drainage networks are functioning.

---

[2] *Etrap* is an administrative district in Turkmenistan.

Finally, within the Karakum water management district, the major source of water is the Karakumderya River, augmented by the Murgab, Kushka, Tedjen, and Atrek Rivers, as well as rivers on the northeastern slopes of the Kopetdag Mountains, underground water, and other small sources of water. The hydro-graphic network is represented here by numerous irrigation canals, pumping stations with pressure pumping lines, collector-drainage systems at Djar, Main Murgab, Central Tedjen, Kakin, Main Gyaur, and Ashgabad "Uniting," as well as Eastern Geoktepa collectors and other small conveying structures.

All of these water management districts are united into a single complex by the Amudarya River. At present, water resources management in Turkmenistan is carried out on the basis of a three-level hierarchy system. The key state agency regulating water use is the Ministry of water management of Turkmenistan; within its structure there are provincial water management organizations ("Suvhodjalyk") functioning in five veloyats. There are also etrap departments of water management that are subordinate to veloyat water organizations and function within the administrative borders of etraps. Veloyat organizations "Suvhodjalyk" have, within their structure, units responsible for operation, maintenance and construction, and other service subunits.

Management of the main watercourse in Turkmenistan—the Karakumderya River—is conducted by the "*Karakumderyasuvhodjulyk*" association, which plays a significant role not only at the inter-etrap level, but also at the inter-veloyat level. This association carries out the management of the "Karakumderya" system through nine "districts of operation," located within the command area of the main watercourse and having no administrative subordination to local authorities.

The water management system of Turkmenistan is a federal sector of the economy that ensures that water resources use is in the interests of not only irrigated agriculture but other users as well. Irrigation canals and reservoirs are sources of household drinking water and cultural-domestic water supply services for the population. They are utilized for the watering of animals and pastures, fishery development, industrial water consumption, energy generation and navigation, recreation, sports and amateur fishing, planting of greenery in urban areas and meeting of ecological needs, and in numerous other ways besides. Taking into account the multipurpose nature of water resources and water facilities use, their development must be based on the augmentation of inter-sectoral links and perspectives of economic development in general. Proper consideration should be given to the nonseparable interrelation between water, sanitary health, and the general health of the population, economy, and environment.

The main directions of the National Programme for the period up until 2020 were reviewed at the fourteenth joint-state meeting of the Council of

Elders, People's Council and nation-wide "Galkynysh" movement on August 15, 2003. The Programme stipulates an increase in agricultural production by 17.7 times, along with ensuring high development rates in mining and processing industries, as well as in construction and other sectors of the economy. Development of agriculture is to be directed at meeting the food demands of the population to a full extent, as well as meeting the needs of industry in raw materials and enhancing the export potential of the country.

The increase in agricultural production is to be achieved by virtue of the improvement of land reclamation activities, increased productivity of animal breeding, the application of mineral fertilizers, and the introduction of advanced knowledge and sophisticated technologies in the production processes. Improvement of lands and the provision of irrigated agriculture with water are to be achieved through the construction of the Turkmen Lake, the building of new reservoirs and by augmenting the capacity of existing ones. In addition, a complex of hydro-technical schemes in the "Dustlyk" reservoir will be put into operation, with a total regulated capacity of 1250 million m^3 of water.

The production of wheat will increase by 2.9 in 2020 as compared to 2000; production of cotton by 4.9 times in the same period, cattle heads by 3.1 times, and the number of sheep and goats by 3.6 times. A 70% increase will be ensured owing to improvement of crop capacity and productivity of cattle and poultry. The stipulated amounts of agricultural produce will be reached through the reuse of areas under crops, obtaining 2–3 yields per year for some crops. The State will continue its policy of supporting farmers, and new economic mechanisms will be established based on economic approaches to management.

Land resources structure will be radically changed as a result of new irrigated land development. At present, 90% of land is State property. As a result of market developments, the share of State property will be reduced, with the share of private property increasing accordingly. As stipulated by the National Programme for the period until 2020, the area of irrigated lands in the country will reach 2240 thousand hectares by 2010 and 4000 thousand hectares by 2020.

Agriculture is the main water user in Turkmenistan; around 90% of total water resources are used for this purpose. Water supply to the population, and water use in other sectors of the economy, consume a small proportion of total water resources. Table 3 (below) shows the structure of water resources use by sector of Turkmenistan's economy.

In the last few decades, the water resources of the country have been used to their full capacity and thus further development of water consuming industries, including the expansion of irrigated lands, will be possible only through improving the efficiency of water resources use and introducing return water use and recycling. Consequently, the main direction of Turkmenistan's

TABLE 3. Structure of water use and water diversion throughout Turkmenistan Statistic data (water management) as of 2000

Description of use	Unit	Total
Abstracted from sources of water	Million m^3	24,828
Amount of water used, total	Million m^3	19,128
including:		
Household needs	Million m^3	476
In % of total use		2.5%
Industrial needs	Million m^3	1,427
In % of total use	Million m^3	7.5%
Irrigation	Million m^3	17,110
In % of total use	Million m^3	89%
Other uses	Million m^3	115
In % of total use	Million m^3	1%
Waste waters - conveyed	Million m^3	589
Collector-drainage water—conveyed	Million m^3	5,302

economy in general predetermines water sector development and underlines the immense role that water management of the country must play in achieving the objectives of national programs and strategic plans. It should be stated that Turkmenistan, while addressing the task of achieving far reaching goals of economy development and raising the living standards of the population, is orienting itself towards fully fledged and efficient use of national water resources, as well as its share of transboundary water resources. To that end, unremitting consideration has been given to sustaining natural complexes and ensuring the ecological security of the country.

The current efficiency of the irrigation network constitutes 0.58. According to calculations, in order to get additional volumes of water through a reduction of losses in the irrigation systems, it is necessary to bring this indicator up to 0.67 by 2010 and 0.75 by 2020. To this end, reconstruction of irrigation canals is to be carried out on existing lands with an area of 1838 thousand hectares and there will be development of new irrigated lands with an area of 2162 thousand hectares, equipped with irrigation networks of high technical standards. Most of the existing irrigation canals will be connected and lined with counter-filtration coating. Some of the canals will be constructed in the form of tubes or flumes.

In addition, in order to reduce water losses in irrigation canals, a number of activities will be carried out, including:

- Selection of optimal variants as they apply to crop structures and crop zoning;
- Improvement and large-scale application of on-farm water recycling;

- Maintenance of the good technical condition of the irrigation network through the conducting of timely repairs and cleaning, along with the application of innovative technologies.

The specific water application norm for a complex hectare of irrigated lands, under current crop contents and patterns and efficiency of watering techniques, constitutes 6354 m^3. By optimizing crop content and patterns, increasing the efficiency of watering techniques, and establishing optimal amelioration regimes in irrigated soils, specific water application norms are to be reduced to 4209 m^3/ha by 2020. For this purpose, the following arrangements are envisaged:

- High-quality leveling of irrigated plots;
- Strict observance of scientifically grounded methods and technologies of water application;
- Development and implementation of water saving technology for surface irrigation;
- Water application to fruit trees and vineyards using drip irrigation systems throughout the whole territory (the area of orchards and vineyards fully equipped with drip irrigation systems will reach 30,000 hectares by 2010, 150,000 hectares by 2015, and 284,000 hectares by 2020);
- Achievement of optimal amelioration regimes in irrigated soils, and combination of leaching with pre-sowing watering;
- Selection and introduction of less water-consuming crops.

At present, 14 water reservoirs are operated in Turkmenistan, with a total capacity of 3018 million m^3, which allows the carrying out only of seasonal water flows regulation. To accumulate flood river flows and ensure multiyear regulation, capacity augmentation of existing reservoirs and construction of new reservoirs is planned. Thus, the additional total capacity of water reservoirs will constitute 696 million m^3 by 2005. The total capacity of all reservoirs will be 7238 million m^3 by 2010, 9498 million m^3 by 2015, and 11361 million m^3 by 2020. Starting from 2015, Turkmenistan will be able to provide for a transition to multiyear flow regulation; this will create the possibility of accumulation of surplus water in high water years in order to mitigate water scarcity in low water years.

The Programme of water sector development for the period of 2003–2020 stipulates the implementation of large-scale works with regard to developing collector-drainage systems designed to convey and accumulate collector-drainage waters. The "Turkmen Lake of the Golden Century" will be the key facility of this system. At present, there are independently functioning collector-drainage systems in each veloyat of Turkmenistan, through which collector-drainage waters are conveyed above the cultural zone to depressions in the Karakum desert. In the Lebap veloyat, waters are discharged into the Amudaya River. Under the impact of high temperatures, water in these storages evaporates very quickly, leading to sharp increases in mineralization. Thus, the water becomes unfit for use, land degradation takes place, and desert vegetation vanishes.

A special situation has taken shape in the Dashgouz veloyat, where large interstate collectors are located. The conveying of water discharges along these collectors, in amounts above the designed levels, during leaching/presowing periods has lead to intensive erosion and considerable deformation of collectors, the disruption of bridges, gas pipelines, communication lines and lines of power transmission, and water conveying facilities, and the water logging of drainage systems, thus causing sharp deterioration in the condition of irrigated lands and pastures.

After construction of the "Turkmen Lake of the Golden Century" is completed, many problems of this kind will be resolved. Furthermore, waterbearing zones will arise in the vicinity of the main collectors and outlets (over a total spread of more than 2000 km.). Within these zones, woodbush vegetation and grass species will begin to actively develop, and consequently the forage productivity of pastures will improve. As a result, the cultivation of salt-resistant crops to meet the needs of local population will be possible.

Construction of the "Turkmen Lake" will be accompanied by scientific studies and field surveys. The priority will be given to studying the capability of highly developed water vegetation to absorb organic matter from water and petroleum products, as well as its ability to block suspensions and extract biogenic substances, heavy metals, phenol, pesticides, and radioactive materials. Construction of the "Turkmen Lake of the Golden Century" is to be completed by 2020.

Operation of the "Turkmen Lake" will allow the implementation of the following:

- Prevention of discharge of collector-drainage water to the Amudarya River, leading to improved water quality in the river;
- Concentration of all CDW discharges into a single outflow from irrigated lands of all five velayats in Turkmenistan, and its conveyance to the "Turkmen Lake";
- Return to agricultural use of vast, distant pastures currently flooded with drainage water;
- Reduction of water levels in interstate Ozerny and Darlyk collectors, thus eliminating backwater and ensuring normal operational regime of drainage systems in Dashguz *veloyat*, and also liquidating destruction risks for transport, gas transporting, electricity transmission and communication lines;

- Improved forage resources and natural productivity of desert pastures in command area of main collectors and outlets;
- Utilization of storage capacities of Zengybaba and Karashor in the interests of further fishery development;
- Improved condition of irrigated lands and reduction of annual salt pressure;
- Creation of new jobs.

The implementation of the above listed activities by 2020 will enable the following:

- Due to the increase in efficiency of irrigation systems (reduction of losses), it will be possible to save 6400 million m^3 of water;
- Owing to the reduction of specific irrigation norms per complex hectare of irrigated land (by 35%), it will be possible to save 8600 million m^3 of water;
- It will be possible to achieve collector-drainage water use to the amount of 3500 million m^3 for irrigating agricultural crops.;

TRANSFER OF WATER RESOURCES MANAGEMENT TOWARD BASIN PRINCIPLES

A. A. Djalalov

One of the most high-capacity irrigation networks in the world has been constructed in the Republic of Uzbekistan and is currently in operation, providing all sectors of the national economy with water supply. Agriculture remains the major water user, utilizing more than 80% of all water withdrawals. More than 4.2 million hectares of irrigated lands constitute the agro-economic potential of the republic, and this is an invaluable resource.

Over the last five years, an average of 53.1 billion m^3 of water has been used annually in the republic, including: 50.5 billion m^3 withdrawn from the country's rivers (33.04 billion m^3 from the Amudarya and Syrdarya Rivers), 0.63 billion m^3 from underground sources and 1.95 billion m^3 from return waters. Around 20% of the total volume of water used annually is formed within the boundaries of the republic and 80% of it is provided by inflow from the territory of neighbouring countries.

The pattern of water use by national economy sector is as follows: household needs constitute 2.9 billion m^3 (5.5%); 4.09 billion m^3 (7.7%) goes in generating energy, including 0.124 billion m^3 (0.2%) for irrevocable water consumption; industrial water use accounts for 0.77 billion m^3 (1.5%); fish farming in-stream takes up 0.40 billion m^3 (0.8%); and agricultural use is 48.9 billion m^3 (92%).

The annual population growth rate in the republic constitutes 3,00,000–3,50,000 people; this causes a continuous increase in priority-driven water discharge to meet the demands of communal-household and industrial water use, at the cost of an even greater reduction of water volume for agricultural use allotted out of the total amount of scarce water resources, which are subject to limitation. Average annual agricultural water use for the last five years has the following pattern: during a non-vegetation period—12.8 billion m^3, including leaching—5.7 billion m^3; wetting of lands—1.8 billion m^3; irrigation of winter grain crops, late and early vegetables, melons and gourds, gardens and vineyards—5.3 billion m^3. During a vegetation period, the figure is 36.1 billion m^3, including irrigation of cotton—14.98 billion m^3; rice—1.31 billion m^3; grain crops—3.15 billion m^3; other cultures—16.7 billion m^3.

The volume of water use in irrigated agriculture is determined by natural-economic conditions, the technical state of the irrigation systems and the methods of water application. A sharp deterioration of water quality has occurred in

P. Wouters et al. (eds.), Implementing Integrated Water Resources Management in Central Asia, 157–166.

all river flows in the region, especially in the Syrdarya and Amudarya Rivers, in all reaches of the channel from the riverhead down to the mouth. Mineralization of water has increased by 0.2–0.3 g/l in upstream areas, by 0.5–0.7 g/l in middle reaches, and by 1.0–1.5 g/l downstream. These increases lead to a build up of salinity on irrigated lands and an increase in water demands for leaching.

Specific annual water use per hectare constitutes 11.2 thousand m^3 on average for the last five years, out of which 8.72 thousand m^3/ha is the figure for a vegetation period and 2.48 thousand m^3/ha for a non-vegetation period. Water withdrawals in the Karakalpakstan and Khorezm provinces constitute 14.3–14.9 thousand m^3/ha due to the vast areas that are under rice.

The total length of on-farm irrigative network is 27 619.7 km, and of inter-farm network 167 378.8 km. The technological level of irrigation systems in Uzbekistan as a whole is low—62% of inter-farm canals and 79.5% of on-farm canals are unlined. The efficiency factor is, on average, 0.86 as applied to inter-farm canals, and 0.75 in on-farm networks (in "new" zones of irrigation in the Syrdarya and Dzhizak provinces this factor is 0.89–0.90). The efficiency factor of irrigation systems throughout Uzbekistan is, on average, 0.64. Major levelling of fields has not been performed anywhere for more than 10 years, and the level and unevenness marks difference is of +20 −30 cm, which is beyond the standard of between +3 +10. Generally, strip and furrow irrigation is used. Non-uniform soil moistening and water losses occur due to poor levelling of fields. Efficiency of water application technique is 0.59–0.70.

More than 75% (3.2 million ha) of the total irrigated area (4.2 million ha) requires the construction of drainage. 2.8 million hectares of this area is equipped with collector-drainage networks, including 580 000 hectares with subsurface horizontal drainage and 413 000 hectares with vertical drainage. Construction of drainage is necessary in an area of more than 400 000 hectares. 136 700 km of drainage networks have been constructed on lands equipped with drainage, out of which 29 000 km are main and inter-farm collectors, and 107 700 km are on-farm drainage networks (including 39 200 km of subsurface horizontal drainage). O&M of 9210 wells is funded by MAWM, including 4214 wells of vertical drainage and 4996 wells for irrigation.

Because of physical-chemical sealing of filters, specific discharge of vertical drainage wells has decreased by 37–79%. In addition, due to the malfunction of a significant portion of wells and pumps, as well as the deterioration of the material and technical supply, the efficiency factor of systems performance has sharply decreased recently and at present does not exceed an average of 0.27–0.34 ha/year, against the recommended standard of 0.60–0.65 ha/year.

Because of economic and financial difficulties, clean out of surface (CDS) is conducted at the level of 50–70% of the standard. The annual need for repairs of subsurface horizontal drainage amounts to 3,000–3,500 km, but only 1,000–1,200 km are actually repaired every year. The major reasons for this are the weak economic condition of farms and the shortage of drain cleaning equipment.

Due to the above stated circumstances, a tendency toward expansion of salt-affected lands has taken shape during recent years (20,87,600 ha in 1991, and 23,31,900 ha in 2001). The area with the depth of ground waters table down to 2 m constitutes 12,67,600 ha. Given the deficit of water resources and limited water use, the outflow of drainage wastewaters throughout Uzbekistan constitutes 22–23 km^3.

Pumping plants provides water delivery in an area of more than 2.2 million hectares. MAWM is responsible for funding O&M of 1466 pumping plants, where 4967 pump units are installed. The service life of more than 70% of the pumps expired a long time ago. The pressure pipelines are in a similar state. As a whole, the radical reconstruction of all pumping plants systems is required. Ten 145-pump units are used in on-farm irrigation and collector-drainage networks. Due to the rise in the price of electricity, petroleum products and spare parts, and the economic inadequacy of farms, farmers almost everywhere suggest devolving O&M of pump units to the State.

MAWM is responsible for operating 52 reservoirs, 28 of which are off-channel and 24 in-stream reservoirs. Their total design capacity is 17,844 million m^3, and effective volume is 14,581 million m^3. To date, all reservoirs are silted by 20–35%, and the Tuyamuyun reservoir has been silted up to 50%. The reservoir's command irrigation area constitutes 1,530 thousand hectares.

Water facilities are owned by the state, and their maintenance and development are funded out of the State budget. The reduction of budget financing has sped up these facilities' physical depreciation and moral obsolescence. The physical depreciation of assets exceeds rates of their production by 2.6–3.3 times. To maintain their working capacity, significant financial and material resources are required.

Water resources allocation, use and control in the republic are subject to regulation by the following documents: the Constitution of the Republic of Uzbekistan, adopted on December 8, 1992; the law *On water and water use* of the Republic of Uzbekistan, May 6, 1993; the decree of the Cabinet of Ministers of the Republic of Uzbekistan *On limited water use* № 385, August 3, 1993; and the decree of the Cabinet of Ministers of the Republic of Uzbekistan *On approval of Regulations on water protection zones in reservoirs and other water bodies, rivers, main canals and collectors, as well as other water sources in the Republic of Uzbekistan* № 174, April 7, 1992.

The limits on water resources use are set in the following order:

The Ministry of Agriculture and Water Resources: by sectors of national economy; by provinces; by sources and volumes of water withdrawals from the Syrdarya and Amudarya Rivers; by main channels.

Province agricultural and water management authority: by inter-district canals; by districts; by internal rivers.
District agricultural and water management authority by each water user.

The following order of allocating quotas for water withdrawal and water delivery to every water user has been established: availability of a water use plan; registered and equipped by water metering devices water outlet; a license for special water use. The set water limit/ quota for water users and water organizations delivering water is the basis for the conclusion of an appropriate contract.

At present, water resources management in Uzbekistan is carried out in an integrated manner according to territorial administrative principles. The Ministry of AWM, as a state agency, is legislatively charged with water resources management. Further on, administrative functions are passed to provincial water authorities, to which inter-district canals and district water authorities are subordinate. Apart from this, an operational regime of more than 50 reservoirs is managed, depending on capacity and significance, by a republican or provincial body.

All interaction with water users is carried out at district level. In particular, the governmental decree of 1993 determined all inter-relations between water users and water management agencies in the Republic of Uzbekistan: water use quotas, the drawing up of contracts, the issuing of licenses for special water use, and the delivery-acceptance of water with appropriate accounting and registration.

After the disintegration of the Soviet Union, the Amudarya and Syrdarya Rivers became interstate water sources. The independent states of Central Asia were faced with the necessity of changing the previous approach to interstate water resources management which had been adopted earlier by the former Soviet Union.

In this connection the joint declaration was adopted in 1991 by five states in Central Asia, and the Interstate Commission for Water Coordination was established in 1992, with two executive bodies: BWO "Syrdarya" and BWO "Amudarya." The management of water facilities relevant to interstate water resources management and distribution was turned over to them on a temporary basis. This agreement was ratified in each republic by a special governmental document. According to the Agreement, interstate water resources management is carried out on the basis of the current structures and principles of allocation, and the normative act in force regulating allocation of resources of interstate water sources.

As we saw, the river basin *Plan of integrated water resources use and protection* is the main normative act that determines the water allocation. In addition, all protocol decisions and mutual obligations pertinent to the given issue that had been adopted earlier remain in force. Every year, ICWC establishes water withdrawal quotas for Central Asian countries from the interstate sources—the Amudarya and Syrdarya Rivers—within the limits set in *Planned volumes*, taking into account available water resources, hydro-meteorological services forecasts, and the general water situation. In accordance with these limits, each republic allocates limits by canals, provinces and districts, taking into consideration specific features of the current water situation. BWO has been given the right to adjust discharges of water intakes as agreed with provincial water authorities within the limits of ±10%, depending on the current water/hydrological situation and weather conditions.

The experience accumulated by ICWC and BWOs in the field of interstate water resources management has shown the virtue of the chosen way as a whole. All issues that arise are addressed and adjusted through meetings of experts and, if necessary, heads of MAWMs of the pertinent country, acting always on the principles of the accepted Agreements.

Now the third stage of reforms to market relations in the agricultural sector is underway in the republic. This involves an intensive re-structuring of agricultural enterprises–private farms are being formed on the basis of former collectives and cooperative farms. Now the number of private farms has reached 90,000, whereas last year there were just 55,000 such operations. The process of transition to private farms is becoming more and more active with every passing year.

The process of re-structuring in agriculture has predetermined reforms in the water sector. The process of reforming the water sector is carried out along three key directions. The first direction is the improvement of new, on-farm water management systems. This has become more complex due to the sharp increase in the number of water users, leading to the need for the introduction of a new system of water accounting. The need emerged to establish appropriate structures charged with on-farm water resources management and distribution control. The optimum way to solve this problem is by establishing Water Users Associations (WUA). WUAs are created on systems of canals, and to date their number has reached 268. The process of WUA creation is very complex; it involves maintaining a minimum number of core staff members and ensuring proper conditions for their performance. At present, WUAs unite 30,300 agricultural enterprises, including 21.3 farms on an area of 6,68,000 hectares.

The second direction is the reform of general water resources management in the Republic of Uzbekistan for the purposes of interlinking WUAs with the top levels of the water management hierarchy. According to the July 2003 Decree of the Cabinet of Ministers of the Republic of Uzbekistan (№ 320) '*On improvement of water resources management*', the transition to hydrographic principles of water resources management has started. This principle is now accepted all over the world. 10 basin organizations responsible for water resources management are being created. They will be organized in river basins and irrigation systems, replacing management based on provincial and district approaches. Such a principle of management has several benefits: it allows planning and use water resources on a scientifically substantiated basis; it results in reduced losses from organizational failures; it increases the efficiency of water use machinery and equipment; it excludes incompetent intervention in water resources management; it ensures the purposeful and rational use of allocated funding; it ensures equitable water distribution and supply according to contractual obligations.

The third direction is the modernization of water facilities, which were constructed 50–60 years ago and have become physically and morally obsolete. Maintaining the technical working-capacity of pumping plant stations constructed in 1970's and 1980's is an especially difficult task. These stations do not meet the modern requirements as to the material and technical basis, the manufacturing of the basic units and parts, and the automated control systems.

To improve water resources management both in the Aral Sea basin in general and within the country as a whole, Uzbekistan participates in the development of various regional and national projects financed by the international donor community. For example, PBAM-1 Project 1.1., *The general water distribution strategy. The Aral Sea basin water resources rational use and protection*, was implemented in 1995 and the *Water resources management and agricultural manufacture in the Central Asian republics (WARMAP)* project, under the TACIS programme, was implemented in 1996. Within the framework of these projects, integrated generalization was carried out as applied to available materials on various aspects of water resources management by sector of the economy, the tendencies were revealed and directions for further development (basic provisions of water strategy) were agreed at that stage of the research.

In order to guarantee provision of a water delivery system to private farms, the decision was taken to launch the stage-by-stage reconstruction of pumping plants, hydraulic structures, reservoirs, irrigation canals and drainage systems. In this connection, feasibility studies have been completed and are in the process of being developed under the following projects:

The Uzbekistan Drainage Project, 1997–1999 World Bank, consortium *Mott Mac Donald-Temelsu-Uzgipromeliovodkhoz* (MMTU).

The Project of agricultural development in Ak-Altin district. Detailed amount of works on improving irrigation and drainage systems. DG Agroprogress International. Germany. 1997–1998.

The Main South-Karakalpak Collector, Joint ecological program JEP-04, TACIS/World Bank, 2001–2003, firm ERM.

The Uzbekistan Drainage Project, Project of drainage outflow diversion in Southern Karakalpakstan, 2002 World Bank, consortium *Mott Mac Donald-Temelsu-Uzgipromeliovodkhoz* (MMTU).

The *Rehabilitation of the Karshi cascade of pumping plants* project, WB, 2002.

The *Water resources management in the Amu-Zang Canal Zone* project; rehabilitation has been planned both at pumping plants and large irrigation canals, 2003. ADB.

The *Rehabilitation of pumping plants of the Amu-Bukhara canal* project, 2003, TDA (USA).

Furthermore, a number of pilot projects have been implemented and are under way aimed at the adjustment of a number of issues emerging in new conditions of re-structuring agriculture and water management, as applied to on-farm irrigation and management:

The on-farm irrigation and management Project. Phase I. TACIS. SOGREAH-SODETEG consortium. 1998–2000.

The on-farm irrigation and management Project. Phase II. TACIS. 2000-2001.

The project *Management of soil and water resources at the level of farms for creating sustainable agricultural systems in Central Asia.* ICARDA, ADB. 1999-2002.

The improvement of natural resources management in Central Asia Project in accordance with the pilot program of improving water resources management at the on-farm level. USAID, PA Consulting. 2001–2003.

The GEF Agency and IFAS carried out the Project *Water resources and environment management. Sub-component A-1—Water resources and salts management* for the purpose of developing regional and national strategies, policies and a program of actions for the period 2001–2003. This project differs qualitatively from the others in terms of the methodology designed for scientifically substantiated decision making in the field of management. The project was based on the analysis and modelling of the current situation, and forecasts of the various development scenarios.

At the same time, numerous mistakes were made in the process of solving tasks at various stages of the project. The drainage infrastructure designed to manage soil salinization and water mineralization processes through land reclamation and agro-technical activities was quite insufficiently analyzed in national and regional scenarios, whilst water demand for the leaching mode of irrigation was not taken into account. The interpretation of high ground water tables on irrigated lands and measures to manage them was carried out without revealing a cause and effect relationship. The positions were wrongly accepted as they apply to conditions of little influence of soil and water salinity impact on land productivity (based on electric conductivity by FAO technique), and do not take into account distribution of salinity relevant to texture of soil patterns and the effect of leaching irrigation on degrees of soil salinity. The issue of collector-drainage water use at locations of their formation and at the regional level has also been studied inadequately.

For this reason, insufficient attention was given to combating soil salinity in water and salt management. Water and salt balances in large reservoirs, in irrigation systems command areas located in various river reaches, and in irrigated zones have remained poorly investigated and forecast. Computer models for calculating water–salt regimes both in separate parts of the river basin, along river channels, and in the basin as a whole, have remained untested and non-calibrated.

Thus each Central Asian country, having obtained independence, started developing economies along their own lines, maintaining their own reformation rates and patterns in the process of transition to a market economy. In these conditions, approaches to complex problems of water resources management, as well as regional and national water policies and mechanisms of their implementation, have not been yet developed; the appropriate legal basis has not been elaborated, nor has a suitable institutional framework, which is able to address the complex problems of water resources management, been put in place.

These are the issues to be tackled in future projects relevant to water resources management in the region:

1. With the purpose of saving water and augmenting land and water resources productivity as applied to each planning zone, it is necessary to determine major measures aimed at combating salinity in irrigated lands and ensuring water saving (leaching and leaching irrigation regimes, improvement of repair works, rehabilitation and construction of irrigation canals, drainage systems, methods and technology of water application, improvement of water use management) as the rational basis for investment in the improvement of regional and national water management.
2. Development of nature protection measures through the minimization of interaction between the river and irrigated areas, and between surface and underground waters.
   In each Central Asian country, there is a need to determine a specific action plan designed to prevent further deterioration of the Aral Sea crisis (relief to the Sea and Priaralie).

3. Development of computer models simulating water management and ecological and economic processes for forecasting outcomes as they apply to planning zones and the river reaches. This will allow identification of trade-offs and mutually beneficial activities.
4. Accurate division of surface, underground and collector-drainage waters into national and trans-boundary water resources (to clearly outline and differentiate international waters and basins and determine their resources; to propose the efficient arrangement of trans-boundary waters management and distribution of pertinent expenses; to develop rules of selection of trans-boundary underground waters in view of their connection with a superficial drain, and the water content mode of the rivers, inflows etc.; to develop rules regulating trans-boundary ground waters withdrawal, taking into account their inter-relations with surface flows and water regimes of the rivers, tributaries, etc.; to develop regulations for monitoring of water quality in trans-boundary sources with various levels of mineralization, in the interests of different water users).
5. Current water-energy relations in the region should be improved. There is a need to work out a more precise and detailed mechanism of interaction between sectors, accompanied by the necessary legal support in the form of special amendments and protocols, or appropriate agreements, which allow the removal of the need for annual agreements. These documents should contain the details reflecting to specific activities in years of various water availability, such as conditions for observance of ecological releases, and share of water for nature, consideration of long-term regulation, the relative shares of the various countries and their obligations as to electricity and fuel, the responsibility for observance of mutual obligations, and an assessment of regional and national projects and works.
6. The institutes charged with basin management should have explicit rules of operation, taking into account all hydrological conditions, distribution of duties for implementing decisions made at all levels, adequate budgets and personnel, access to appropriate data (including the daily discharge and withdrawals), and the means of monitoring the execution of the accepted decisions, including the power to impose sanctions for infringements.
7. Development and interaction of regional and national information systems, providing day-to-day analytical data exchange, especially concerning data on conditions of water scarcity and floods, interlinking data obtained by BWOs, Hydro meteorological Services, and MAWM, especially when immediate information is needed on changes in discharges and water levels in the rivers.

8. Perfection of the organizational structure of water resources management both at regional and national levels.

These issues, as well as additional specific projects aimed at rehabilitation of separate water facilities, were reflected in the proposals in the water related section of PBAM-2, approved at the ICWC session on May 6, 2003 in the city of Alma-Ata (protocol decision).

BASIN MANAGEMENT BASED ON RESOURCE CONSERVATION

Zh. Bekbolotov and A. Jaloobayev
Water Department, Kyrgyz Republic

Many regions in the world face water scarcity. Among these are the countries of Central-Asia, which have limited water resources. For this reason, efficient water use and water conservation are top priorities, especially in terms of irrigated agriculture.

The natural climatic conditions of the Central-Asian region, allied to education and awareness of the subject, mean that traditionally the local population has displayed a careful attitude to water. Unfortunately, with large-scale irrigation development these traditions were somewhat forgotten, a process exacerbated by the birth of a myth of future water abundance from Siberian rivers. The Aral Sea started to shrink, water transfer from Siberia became impossible and the issue of efficient water use and water conservation became critical.

Under water scarcity, the only means of equilibrium between water supply and demand is demand management through water conservation.There is a positive experience in the region, based on the results of the WUFMAS subproject (TACIS WARMAP), sub-component A-2 (Participation in water conservation, 1999–2000); GEF project, IWMI-SIC ICWC project, 'Best practice in water conservation adoption in SyrDarya and AmuDarya basin'. This experience showed that, in the Aral Sea basin, water conservation methods known from science and tradition are being practiced, but not in a widespread manner. Awareness of potential water productivity may allow countries to almost double agricultural production under a water expense decrease of 10%.

As an example, we might turn to data from the WUFMAS project obtained from crop yield monitoring on 22 control fields under typical conditions. According to this data, irrigation water losses at the level 'inlet to field-cotton rooting zone' amount to 51% across the region. The experience of the countries taking 4 t/ha cotton under irrigation of 5 th.m^3/ha says much about the huge potential of water conservation. The following comparisons underline the gap in water conservation in the region. Each Tashkent resident spends nearly 1000 l on water per day, while in Shanghai this figure is 577, in Hong Kong 402 and in Kuala Lumpur just 354.

Many countries fully utilise wastewater. Such wastewater in our region amounts to 6 m^3/yr, and is not used at all, creating a load on the environment. It is important to remember that the water conservation effect is not limited by

P. Wouters et al. (eds.), Implementing Integrated Water Resources Management in Central Asia, 167–177.

irrigation water saving. It has been proved that water saving leads to irrigated agriculture productivity increases. From WUFMAS data on land and water productivity it is evident that, on average, the 'gross-field' irrigation norm for cotton is 7243 m^3/ha, comprising 2039 m^3/ha of leaching and recharge and 5204 m^3/ha of vegetation irrigation. Under an average cotton yield of 2.33 t/ha at field level, average weighted irrigation water expenses per yield unit amount to 3,110 m^3/t, under water use productivity at field level of 0.32 kg/m^3 (indicators may vary within 1600–10,340 m^3/t and 0.1–0.63 kg/m^3 respectively). For winter wheat, the average weighted 'gross-field' irrigation norm amounted to 4575 m^3/ha under an average yield of 2.23 t/ha; irrigation water expenses were 2080 m^3/t under irrigation water use productivity at field level 0.49 kg/m^3 (data deviation from 180–5750 m^3/t and from 0.17–5.65 kg/m^3 respectively).

Top-heavy water supply to the field leads to a reduction in land productivity because, on the one hand, nutrients are washed out from the soil which cannot be adequately compensated for by fertilisers and, on the other hand, because of rising ground water and soil salinisation. According to WUFMAS control field data, mobile phosphorus losses due to its washing out reach 65%, and potassium losses are 50%, whilst soil salinisation reached 51% over the last 2 years.

Comparative analysis of irrigation water productivity based on gross product shows the following: If in Israel water productivity equals 0.52 \$/m^3, in Uzbekistan it is 0.06 \$/m^3, in South-Kazakhstan oblast it is 0.14 \$/$m^3$and in Kyzylordaoblast the figure is less than in Uzbekistan. Although these figures reflect local market competitiveness, they are indicative. It is worth noting that there is a difference in water use productivity for fishing culture between Israel and The Netherlands.

In each planning zone, river site and command irrigated area within the country, and within the basin the following circumstances, indicators and factors should be analysed and evaluated:

- Potential land and water productivity based on available information, especially for dry years;
- Specific water consumption under minimum water discharge for biological production using CROPWAT-FAO methodological approaches;
- Reasons for production shortage (connected to reclamation and water-related factors) and possibilities for their elimination with assessment of priorities and measures undertaken;
- Planning zone water-salt balance based on previous data; possibility of their reduction to parameters providing ecologically sustainable development (minimum salt exchange between river and irrigated area and between aeration zone and ground water with gradual salt storage decrease in aeration zone and planning zone as a whole);

- Possibility of return water maximum involvement and utilisation in place of origin;
- Possibility of released and ground water use;
- Possibility of organisational water losses reduction at all hierarchic levels;
- Unproductive water losses in all irrigation system links, first of all, at irrigated field, assessment of which will allow identification of the least capital consuming measures of water conservation;
- Reduction of return water release to the river and water quality improvement as result of water saving. Specific task is to analyse organisational water losses,
- Identification of mistakes in water allocation and in water resources management, including excessive water losses in channel, water release to natural depressions and bad water use downstream. By the joint efforts of regional and national experts, these losses should be identified and mechanisms for their reduction and prevention should be developed.

The countries located in the flow formation zone are ready to provide to all regional organisations the appropriate information about flow forecasts and climatic indicators. However, for this to occur, the participation of all the countries involved is required in hydro-meteorological activity funding.

Under market conditions, economic mechanisms play an important role; water use efficiency and water saving indicators both at national and international level will depend on water charges and their economic mechanisms will dominate. During the period of market conditions formation in the Kyrgyz Republic, the population, including those in rural areas, clearly understood the necessity of economic reforms. Economic and structural transformations made according to the Constitution are reflected in the water sector. Adopted in 1994, the law 'About Water' approved the basic principle of paid water use, water right distribution and national and international water relations. Along with this, environmental legislation was also being developed. A basis for water tariff policy and public water associations was laid. Last March, Jokorgu Kenesh approved the law 'About water users associations', which defines the objectives, principles and main directions of state policy on further sustainable development of water users associations.

At present there are several different approaches to water charges for irrigation in Central-Asian countries. For example, in Kazakhstan charges for 1000 m^3 water delivery are 14,865 tenghe (0.105 cent/m^3) and charges for 1 m^3 surface water are 3.02 tiin (0.021 cent/m^3); in Kyrgyzstan, meanwhile, water charges are differentiated. During the growing period they are set at 30 som/1000 m^3 (0.063 cent/m^3), whilst during the non-growing period they fall to10 som/1000 m^3 (0.021 cent/m^3). By comparison, in Tajikistan 1 m^3 water costs 0.3 diram (0.107 cent/m^3). Meanwhile, in Turkmenistan and

Uzbekistan there are no water charges for irrigation, although in Uzbekistan a charge is included in land tax and excessive water diversion is also paid for.

Meanwhile, paid water experience in Kyrgyzstan shows the importance of this charge for water conservation. According to SIC ICWC analytical data, based on WARMIS and 'IWRM-Ferghana' project data, since paid water use was introduced in 1995, oblasts of Kyrgyzstan in the Ferghana valley have sharply reduced water diversion for irrigation.

In the period before the introduction of paid water use (1986–1995), annual water diversion was 3.7–4.7 bln.m^3. After its introduction, diversion amounted to 2.67–3.68 bln.m^3 (without taking into account the wet year of 1998). If we consider the total water diversion for irrigation purposes by the three oblasts in the Ferghana valley in 5-year intervals, it can be seen that in 1986–1990 the figure was 22,271 mln.m^3, in 1991–1995 it was 19,655 mln.m^3 and in 1996–2000 (after the introduction of paid water use) it fell to 16,987 mln.m^3. This reduction has happened within the same irrigated area, although under some cropping pattern changes. Thus, paid water use in Kyrgyzstan has facilitated a sharp reduction in water diversion for irrigation purposes.

Paid water use and the establishment of WUA led to a 30% reduction of water diversion for irrigation. Kyrgyzstan does not face water scarcity but it does recognise environmental needs and actively participates in water conservation programmes. Thus the introduction of IWRM principles on the Aravan-Akbura canal. Kyrgyzstan creates demonstration plots and pilot projects, organises training and educates people in water-saving behaviour. Demonstration plots within WARMAP, GEF and IWRM-Ferghana projects and the SPECA programme serve as examples.

World experience shows that the participation of farmers in water resources management, as a rule, creates conditions for more efficient water use.

There are four main principles stimulating concerned parties and water users to participate in institutional reforms, something that will increase the number of stakeholders and give the programme more authority, thus increasing the chance of reform sustainability. Although such principles of joint participation are common for all countries, the mechanism of their implementation may differ depending on the socio-cultural situation in the country concerned. The four principles are as follows:

- sharing information—Access to information for landowners;
- consultation—Discussion of transformation alternatives and priorities approval;
- joint decision making on technical questions;
- giving authority to water users though group responsibility in management.

In conditions of collective and state farms separation, and where water user numbers are increasing, the issue of water productivity in irrigated farming and irrigation network maintenance is solved by the establishment of a WUA. With the establishment of a WUA, farmers and other agricultural organisations have the opportunity for direct participation in water resources management and irrigation system reconstruction. Over the last 7 years in Kazakhstan and Kyrgyzstan, great experience in the establishment of WUAs has been accumulated. Kyrgyzstan is a pioneer in this field, having started down this road in 1995. There are now 316 WUAs in the republic.

One of the most serious shortcomings of national legislations in Central Asia is the lack of, or the weakness in, the legal base in the water conservation field. Existing norms require rational water use but they are declarative. There are no special sections on water conservation, a subject that has not received any methodological, organisational, or technological provision. Water conservation not only covers water deficit but has also an economic character and requires legal provision. It is known that water resources in Central Asia and many other regions are fully developed, and additional resources are absent. Even if they are available, expenses for their development will double every 20 years. Water conservation is the only viable approach for long-term prospects in Central Asia.

Irrational and inefficient water use causes damage to the region's economy, a regional problem exacerbated by trans-boundary waters joint use. To compensate water losses, water supply should be permanently increased with additional charges for electric energy and other resources. If this happens, water diversion and disposal increase substantially, a factor that requires additional expenditure for environmental measures.

For any water-related system, water losses lead to system capacity increase, reliability decrease and water and resources losses increase. Calculations show that water conservation and water losses reduction can reduce costs per 1000 m^3 by \$95–100 in municipal water use and can reduce operational costs by \$35–40/yr per 1000 m^3. Water conservation policy should have an integrated approach covering progressive development, structures construction and modernisation, allowing a high technological level of water conservation. Thus, water conservation policy simultaneously facilitates water deficit compensation and protects water resources and the environment from exhaustion and pollution.

In this regard, energy conservation experience can be very useful. In Kazakhstan, energy conservation policy is based on a special law approved in 1997. This law creates economic and organisational conditions for the effective use of energy and environmental protection. In this legal act, the main principles of state power policy are confirmed and the main directions of energy conservation are defined.

The state regulation system in energy conservation and its economic mechanism are given. The legal norm defines a priority in renewable power resources use, power engineering development programmes and ecological issues solutions. Sections covering the legal mechanism of education, information provision and scientific research are given much importance. Some of the major principles of state policy in energy conservation are detailed below:

- priority of fuel-power resources use efficiency over their mining growth rate, heat and electric energy production;
- priority in health and safety, social conditions provision and environmental protection under fuel-power resources mining, transportation and use;
- state regulation of power conservation area;
- obligatory reliable accounting of fuel-power resources.

In 1998, the law 'About power conservation' was adopted. The goal of this law was the creation of conditions for the increase of fuel-power resources efficiency in terms of mining, storage, transportation, processing, distribution and use and the protection of consumer's and producer's interests through the regulation of relations between the state and juridical and other persons involved in the energy conservation field. State energy conservation policy is ensured by:

- the creation of economic and legal conditions to stimulate juridical and natural persons toward energy conservation on the basis of a combination of consumers, suppliers and producers interests stimulation, including techniques and technologies of effective power resources use, accountability and control over power resources spending;
- the development and implementation of state projects and programmes on energy, alternative fuels and renewable and secondary power resources conservation and use;
- the implementation of demonstration projects of high power efficiency;
- the implementation of economic, information, educational and other directions in energy conservation area;
- international cooperation in power resources-use efficiency.

In the Kazakhstan law 'About energy conservation', a separate chapter (Chapter 12) is devoted to the stimulation of juridical and natural persons for power conservation measures (Chapter 12):

1. Those juridical and natural persons who achieve power resources spending reductions of 1% from the established limit have a right to a tariff reduction of 1.5%. Total tariff reduction should not exceed 25%.

2. Juridical and natural persons using unconventional and renewable resources for more than 25% of their total power may have their tariff reduced by 50%. Compensation for reduction of prices and tariffs (items 1 and 2 above) is drawn from territorial funds of energy conservation after power expertise.
3. For rural areas and areas with a complex ecological situation, 80% of equipment and structures for unconventional and renewable resources is compensated from the national and local budgets. The list of eligible regions is defined by the Kazakh Government on the basis of documents submitted by local authorities, power supervision and environmental bodies.
4. Juridical and natural persons developing and constructing power installations and units have tax privileges defined by the Tax Code of the Republic of Kazakhstan.
5. Juridical and natural persons saving fuel or energy also have tax privileges defined by the Tax Code of the Republic of Kazakhstan.

In the Kyrgyz law 'About power conservation' there are also incentives, but tax privileges themselves are not foreseen. Chapter 23, entitled 'Stimulation of power conservation' lays out these incentives:

- grant allocation from the fund of power conservation for socially meaningful projects, maximum share of which in total funding is defined by legislation of Kyrgyz Republic;
- compensation in power resources prices of cost of energy of conservation technologies, including capital cost of power installations using renewable resources constructed within framework of state programmes of power consumption efficiency increase;
- reimbursement of expenditures paid by suppliers for resources under-use by consumers because of power conservation;
- setting of accelerated depreciation norms on power conservation equipment, devices, materials, control, regulation and account devices, list of which is defined by the state body of power conservation management;
- State support for power conservation projects and programmes;
- provision of government guarantees to foreign investors financing power conservation projects, within financial means foreseen by state budget for power conservation measures.

In Tajikistan there is a law 'About power conservation' and a Government provision 'About measures on introduction norms in heat and power use in the national economy of Tajikistan'. This law provides a basis for power conservation policy.

In Turkmenistan there is no special law regarding power conservation. Moreover, this issue is not considered as a priority because there is no restriction on power resources use in the country and some resources are supplied to the population free of charge.

In Uzbekistan, in 1997, the government approved a power strategy, and in May 1997 the law 'About rational power use' was approved. This law includes power consumers, producers and suppliers. It foresees the establishment of standards in power conservation, power certification, accounting and auditing.

Based on power engineering experience, it would be expedient to create a legal base for power conservation in all countries of the region. In this law, tax and custom privileges measures aimed at power conservation, new technologies and the introduction of equipment, as well as water saving incentives, should be included.

It is desirable to create a favourable investment climate both in each country and in the region as a whole, promoting investment in irrigated agriculture efficiency, including water conservation. The states of the region carry out a policy of grain independence. In this connection, areas under wheat are expanded at the expense of cotton forage crops and melons. As a result, livestock productivity is reduced because of a lack of forage base. Vegetables, fruits and melons production satisfies the local market and also allows for some export. The actual level of exports does not correspond to the full potential because of imposed restrictions. As a result, agricultural producers lose potential income.

Food imports are reducing both in volume and variety. Government policy is directed at further import reduction and food export increase, and does not foresee regional specialization. Serious gaps in food production growth compared with population growth leads to food problem aggravation. The population's access to adequate nutrition will depend on agricultural production efficiency and economic development in general. Calories are the main indicator of nutrition. It is estimated that the number of calories consumed per capita should reach 2700–3200. This requires appropriate agricultural production. To this end, in the near future measures to strengthen the agricultural sector, expansion of irrigated areas and new water resources development are needed, given that water is a main factor of irrigated farming.

Irrigation technique perfection and crop yield increase will arise from the following. Firstly, irrigated farming needs intensification in order to satisfy population demand for food. Secondly, water spending for irrigation should be evaluated with regard to future water conservation technologies. At the same time, it is necessary to take into account traditional water conservation methods. One of the ways this can be done is through field protecting forest stripes, which can save 15–25% of water.

Presently, in the tropical belt, there is a spread of Mexican wheat whose potential yield is 10 t/ha. This crop served as the basis for the 'green revolution' in India and some other countries. It is evident that without the help of technical measures it is impossible to solve the world's food problem.

Biology plays an important, and perhaps even leading role. With help of biotechnologies, combined with advanced irrigation techniques (drip and other in-soil irrigation), it is possible to achieve desirable results, including water resources provision. These resources should be renewable up to some limit, which depends on the diversion volume and water quality connected with its use.

Food production growth will be ensured by yield increase .The irrigation norm will reduce due to improved irrigation techniques and crop selection. Total irrevocable water expenses will grow insignificantly. This indicator must be stabilized and the main effect will be decreased water expenses per unit of production. That is why the main strategic direction in water sector development is water expenses per production unit decrease. These measures should be carried out in all economic sectors, but in the first place in irrigated agriculture.

Integration processes between countries in the region, based on understanding by governments of the mutual benefits to be had from cooperation, are very important. Taking into account the long-term forecast for grain price increases in the future, the issue of grain shortages can be solved by grain production in Kazakhstan. It is known that wheat cultivation on irrigated lands is unprofitable. That is why arguments for wheat growing are unconvincing. Wheat grown on non-irrigated land in the Kazakhstan steppes gives much better results in yield and profitability. By carrying out a moderate policy in grain self-sufficiency on the basis of interstate cooperation, it is possible to:

- Increase areas under forage crops and, especially, alfalfa;
- Restore crop rotation using alfalfa as a nitrogen fixing crop for soil fertility rehabilitation;
- Increase production of forage grain;
- Increase livestock forage base efficiency;
- Increase livestock productivity.

Regional cooperation should be based on the economic benefit from agricultural production, and should cover such spheres as milk and meat production in Kazakhstan and Kyrgyzstan, livestock development and sugar beet cultivation in Kyrgyzstan, vegetable cultivation in Turkmenistan and Uzbekistan and fruits and melon crops in Tajikistan, Turkmenistan and Uzbekistan. Greenhouses and film-cover technologies, barriers processing and storage and dry fruits production can facilitate export potential in agricultural production. Regional cooperation will lead to food import reduction, leaving a negligible volume of export within the region in accordance with economic expediency.

It is clear that no one country could agree with a policy of dependence on food imports until a period of political and military security is achieved. Specialists predict urban population growth, with correspondent available water resources redistribution. Urban and industrial water supply will become the highest priority. The agricultural sector will shift toward the production of valuable crops. As world experience shows, in spite of potential benefits, water will be redistributed from the agricultural sector to industrial needs. In countries such as Israel, Cyprus and Malta, governments have successfully shifted the population away from agriculture to industry, commerce and tourism. Agriculture is limited to valuable export crops and food for the population is bought, but not produced locally. A strategy of irrigated farming production change toward food import, paid for by urban growth and commerce development (the so-called import of 'virtual water') is being realised. In our region, the necessity and importance of 'virtual water' (water used for another purpose) for regional food provision should be recognised. The main parameters of food provision, in order to prevent possible crisis, should be based on regional cooperation and integration in agricultural and food production.

Conclusions and Recommendations

The most important condition for water protection and water conservation policy is a flexible and appropriate legal base, corresponding to international practice. As was mentioned before, one of the serious shortcomings of national legislations in Central-Asian countries is the lack of, or weakness in, the legal base regarding water conservation. A lack of water policy and water protection and conservation legislation can lead to water demand growth. Taking into account the above permanent legislation, perfection is necessary at the national level, along with coordination of national legislations at the regional level and regional conventions development. For maximum effect in water conservation policy, international cooperation is very important. A good base for this could be the development of the model law 'About water conservation' for all CAR countries.

In the economies of the countries involved, structural transformations are underway. Land ownership is changing and investment redistribution is taking place. The high price of power is preventing the achievement of the maximum benefit from available resources, which in turn leads to investment activity intensification in the water sector. With respect to these circumstances, the strategic goal of national policy in water resources is the realisation of long-term measures directed at eliminating the negative consequences of limited water resources and creating conditions for economic growth, social and ecological issues solution and interstate water relations regulation. In this respect, it is necessary to understand that water is an economically valuable resource that determines the country's development sustainability, and water quality issues should therefore be considered alongside issues of water quantity.

An integrated approach to water resources management and water conservation, water pollution reduction and water diversion from natural sources, as well as economic regulation of water use on the basis of a balanced tariff system, are the main principles of water policy. Reduction of water diversion from natural sources should be considered as the most important aspect for the ecological security, rehabilitation and preservation of rivers. Huge multi-fold effects are achieved in all economic spheres related to water. Water diversion reduction leads to minimisation of expenses for water object construction, waste reduction and a reduction of the load on the environment. It is important that a source fully retains its functional element as an environmental component. Independently, water policy, level and goals should lead to water conservation as a result of consumption reduction and prevention of negative ecological impacts.

The participation of the population in the decision-making process links water users and administrators. Also of great importance is the establishment of WUAs, and their strengthening role as a tool of water policy realisation at public and private levels of water management.

All of the above-mentioned factors should be realised within a framework of common water conservation policy and strategy, coordinated with all the countries of the region.

Printed in the United States
75566LV00002B/331-351

9 781402 057304